ENVIRONMENTAL RISK MANAGEMENT

ENVIRONMENTAL RISK MANAGEMENT

Paul Pritchard

Routledge
Taylor & Francis Group
LONDON AND NEW YORK

earthscan
from Routledge

First published in the UK and USA in 2000 by Earthscan

For a full list of publications please contact:
Earthscan
2 Park Square, Milton Park, Abingdon, Oxfordshire OX14 4RN
711 Third Avenue, New York, NY 10017, USA

First issued in hardback 2017

Earthscan is an imprint of the Taylor & Francis Group, an informa business

ISBN 13: 978-1-138-16768-1 (hbk)
ISBN 13: 978-1-85383-598-8 (pbk)

Typesetting by Paul Pritchard
Cover design by Andrew Corbett

A catalogue record for this book is available from the British Library

TABLE OF CONTENTS

List of Tables and Figures *vi*
Acknowledgements *vii*
Executive Summary *viii*
About the Author *ix*

1 Introduction 1

2 The Nature of Environmental Risks 7

3 Property Related Environmental Risks 15

4 Financial Risk Transfer 42

5 Risk Management Integration 50

6 Decision Making with Uncertainty 72

7 Risk Assessment and Management Frameworks 82

Acronyms and Abbreviations *102*
References *104*

Appendix A – Environmental Issues in Corporate Transactions *110*

LIST OF TABLES AND FIGURES

Tables

2.1 Environmental Accidents 13

3.1 Uses of Land Which May Give Rise to Contamination 16
3.2 Environmental Risk Factors 17
3.3 UK Incidents and Contaminated Land Sites 34
3.4 Contractual Issues Requiring Allocation of Responsibility and Liability on Contaminated Land Projects 36
3.5 Summary of Remediation Techniques 38

5.1 Activity Assessment 63
5.2 HSE Financial Losses 69

6.1 Remedial Action Options 75
6.2 Property Acquisition – Costs 79
6.3 Property Acquisition – Costs with Regret 79

7.1 UST Release Potential Factors 90
7.2 UST Damage Potential Factors 90
7.3 Environmental Risk Ranking – Site Risk Factors 92
7.4 Dutch Intervention Values 97

Figures

3.1 Property Risk Management Process 20
3.2 RICS Property Observation Checklist 21

4.1 Contaminated Land Remediation Costs 47

5.1 Risk Management Process 51
5.2 EBRD Review Memorandum 57
5.3 Ranking Matrix 58

6.1 Landfill Leachate Migration 74
6.2 Remediation Decision Tree 76
6.3 Fuzzy Sets – Tall People 80

7.1 OPRA – Increasing Risk 87
7.2 Risk Rating Chart 91

ACKNOWLEDGEMENTS

My thanks are extended to colleagues and friends at Royal & SunAlliance who have helped me in the preparation of this publication: Phil Bell, Karen Jeffery, Julia Graham, Linda Lancaster and David Liptak. Last, but most definitely not least, I would like to acknowledge the continuing support of my wife Charlotte and my parents Dorothy and Bill.

EXECUTIVE SUMMARY

This publication considers recent developments in environmental risk management as they relate to commercial organisations, including risk transfer through insurance. It starts by looking at the characterisation of risks based on the hazard-pathway-receptor principles, emphasising the importance of site specific factors.

A tiered approach to the assessment and management of property related risks is discussed, building on existing guidance such as that available from the Royal Institution of Chartered Surveyors (RICS). Research on the outcomes of environmental pre-acquisition studies complements consideration of contaminated land and fire related risks.

The increasing use of risk-based approaches within business functions such as accountancy (as well as risk management) offers the potential for environmental risks to be considered as a part of strategic control assessment. This might be regarded as a development of the ongoing integration of the environment with functions such as quality or health and safety. Checklists and case studies are presented to assist in the review and assessment of environmental risks.

The potential for tools such as Monte Carlo analysis and fuzzy logic to assist decision making under uncertainty is considered, leading on to a review of risk assessment and management frameworks. These include the Hazard Ranking System (HRS) and Risk Assessment Guidance for Superfund (RAGS) which underpin the US Superfund programme, through to models for ranking landfill sites or underground storage tanks. Methodologies such as Operator and Pollution Risk Appraisal (OPRA) underline the increasing use of risk-based models by regulators, in this case helping to focus inspection resources.

ABOUT THE AUTHOR

Following completion of research in environmental chemistry Paul Pritchard worked at the Laboratory of the Government Chemist investigating soil and water contamination. Several years in environmental consultancy saw a shift towards risk management work. He has worked in the Royal & SunAlliance Group since 1994, with responsibilities including the assessment of underwriting environmental risks and the implementation of internal controls.

In his spare time he tries hard to reconcile a concern for the environment with an interest in classic motorcycles.

CHAPTER 1
INTRODUCTION

This publication started from the desire to provide an update to *Managing Environmental Risks and Liabilities.* In the five years since it appeared there has been a range of significant developments in the field of environmental risk assessment and management. There have been major changes in the risk-based regimes for dealing with contaminated land in both the Netherlands and the UK and environmental risk management standards have appeared from the US. At least three significant trends are apparent. The greater use of risk-based methodologies in regulation has been complemented by the application of risk management ideas in integrative initiatives within organisations. Finally, there has been increasing pressure for communication and transparency, reflecting societal desires for reassurance on the use of appropriate risk controls in organisations.

The evolving interest in environmental risk assessment has seen the discipline increasingly extended from the academic sphere to become important in government strategy as well as in commercial organisations. This is largely due to its potential for facilitating decision making in a number of areas that present particular difficulties. For example, there may be inadequate data, or what data is available is subject to considerable or unknown uncertainty. Cynics might also argue that it adds a veneer of scientific credibility to aid politically motivated decisions.

Another advantage associated with an environmental risk-based approach in a commercial context is that it closely follows established business thinking. In this way it can be ensured that environmental issues are given equal prominence to, and more easily integrated with, other business functions. For example, organisations are already likely to have a risk management function to identify business risks, allocate resources and pursue risk reduction. Contingency planning would also be important in both environmental and general business terms. In addition, it can reflect uncertainties associated with the assessment and significance of environmental impacts in a consistent and meaningful way.

Risk and Legislation

The fundamental importance of legislation as the baseline for environmental risk management cannot be overstated. However this publication supports the idea that there is considerable merit in moving beyond a compliance focused approach. Future developments in the legal regime are also likely to require more detailed investigation of issues such as the current extent of land contamination, that are amenable to a risk-based approach. For example, in the implementation of the EC Directive on Integrated Pollution Prevention and Control (IPPC) it is likely that the UK government will require the operators of IPPC processes to report initially on the state of their site. Prior to closure of the site a clean up routine could then be employed to either the IPPC regime or the new contaminated land regime (eg for pre-existing non-IPPC contamination) or a combination of the two (DETR, 1998). In other areas such as climate change, which might profoundly influence future organisational decisions, there is currently little in the way of legal requirements (eg in energy efficiency or choice of fuels) although this is likely to change with the forthcoming IPPC regime. Even with relatively few current 'legal drivers', important issues such as this are likely to form part of the risk agenda.

Risk management programmes are therefore likely to go beyond the requirements of legislation and may even pressure regulators into the provision of further guidance and standards in environmental areas. For example, any assessment of property transfer risks that did not include a recognition of the potential of underground storage tanks to cause significant environmental damage would be seen as incomplete, even though it is an area that is currently not the subject of strong environmental regulation in many countries. In the case of contaminated land, pressure has come in the UK from risk managers on to legislators to provide clear guidance on the standards that go beyond existing guidelines (eg for clean-up purposes).

Risk and Regulators

On the regulatory front the strategic utility of a risk-based approach has been recognised within the Environment Agency by the establishment of a National Centre for Risk Analysis and Options Appraisal. This multidisciplinary body is involved in the development of risk-assessment techniques, forecasting the future state of the environment for scenario analysis and providing guidance on costs and benefits to the Agency.

Risk assessments may also be applied as an operational tool. In determining whether to serve a works notice for ongoing or likely pollution under the Anti-Pollution Works Regulations (1999), the Environment Agency will carry out a risk assessment considering the nature and quantity of the material, the likelihood of it entering the waters in question and the quality of the water body. Models such as the Operator and Pollution Risk Appraisal (OPRA) have been developed that can help to focus inspection effort.

In explaining why risk assessment is important for regulators '... *in a society where a primary role seen by the electorate for government is to protect against risks ranging from collapsing buildings, sinking ferries, environmental disasters and even the minutest threat to health*' the Parliamentary Office of Science and Technology (POST, 1996) indicated that risk assessment is an attempt to bring consistency to a complex range of pressures to achieve:

- equity – all having similar rights of protection
- value for money – allocating finite resources according to their relative importance

Significantly, there has been the increasing involvement of the general public in risk-based environmental issues from questions about the real risks posed by BSE through to recent concerns about endocrine disruptors and genetically modified food. A simple affirmative response to the question – is it safe? – has proved elusive in the face of risk and uncertainty about outcomes. Complementing this greater desire for information and openness has been an increase in the willingness to challenge the views and opinions of experts.

A recent review of various environmental risk management practices at a national/regulatory level has highlighted these issues (Power and McCarty, 1998). This showed a move towards greater emphasis on balancing multiple social goals in an attempt to achieve consensus. In turn this has meant a lesser emphasis on quantitative risk estimates and is leading towards more inclusive processes where those affected by risks become involved in the decision making process. Moreover, it is becoming recognised that the availability of quantitative risk estimates does not necessarily signpost the best risk management option, particularly where there is considerable uncertainty about outcomes.

The Confederation of British Industry (CBI) has recently produced a discussion document on improving environmental regulation (CBI, 1998) based on a survey of members. It stated that risk assessment is a key tool in achieving sustainable development as it requires business and regulators to assess tolerable risk against the cost of reducing it. It recommended a closer alignment of health and safety and environment regulation with a recommended move towards the "goal-setting, risk-based approach" of health and safety law.

The CBI discussion paper highlighted the potential for risk assessment to stimulate debate. An editorial in the journal (ENDS, 1998a) titled 'A blast from the past by the CBI' suggested that the CBI was embarking on an anti-regulation phase and that the Health and Safety Executive (HSE) was not the appropriate model for environmental regulation. This view was based on their opinion that the HSE was inward looking and that open, participatory forms of decision making would be needed in future environmental decision making.

Integration of Management

There has been a marked trend towards the integration of environmental controls within organisations. Primarily this has been with related disciplines such as health and safety and quality (here reflecting the systems-based nature of standards developed). The CBI survey referenced above noted the desire for greater integration at an operational level, while cautioning that the strategic aims of health and safety and environment could differ and even contradict.

These developments must clearly be relevant in any discussion on environmental risks, however, it is apparent that developments in the broader commercial sphere have become more relevant. Corporate functions such as treasury, internal audit, property and risk management have been moving towards a broader range of responsibilities including environmental concerns. Risk-based approaches have generally been adopted, which are in many cases entirely compatible with their purely environmental counterparts.

The expanding role of risk managers (historically cast in the role of insurance buyers) has been described by Barker (1997). They are moving from insurance related issues such as property and business interruption to a broader view of corporate risk assessment and control, with corporate executives needing to think of risk management in terms of ongoing activities and controls as well as financial performance. She notes that *'Risk managers will increasingly work with their Environmental Managers and consultants in a co-ordinated way, reaping real rewards in ensuring not only real reductions of emissions, spillages and pollution incidents but also identifying energy saving and waste reduction opportunities'*.

Risk profiling processes are now being introduced to identify key current and future risks to the organisation that could include environmental concerns. This process starts from the viewpoint that as a baseline, legal compliance must be assured but then moves further to anticipate forthcoming legislation, accident risks, reputational risks etc. In this context environmental issues have been seen to offer opportunities for, among others, the cost-efficient use of resources and enhanced stakeholder value (including investors and the general public image).
This coincides with the view from treasurers who indicated that an approach to risk control where line management focuses on operational risks, the board monitoring strategic items, the treasurer

focusing on financial risk and the risk manager considering insurable risks would potentially fail the needs of businesses in the twenty-first century (Harris-Jones, 1998). A number of generalised business risk models have been developed to enable organisations to address the strategic and cross-functional nature of this issue.

Communication

The communication issue was highlighted in a US context (Tusa, 1994) where a consistent definition of environmental risk management was seen as crucial to achieving consensus on environmental issues. Areas of historical inconsistency were given as:

- differing health risk assessment methodologies being used by regulators – meaning that regulations might utilise standards based on either available technology or risk reduction, with or without regard to cost (meaning the cost of risk reduction varies greatly);

- conservative assumptions made in risk estimation (particularly in Superfund) resulting in expensive remedies with limited risk reduction;

- media failure to communicate environmental risks in perspective; and

- poor understanding by senior executives of environmental issues, and the focus of some organisations on regulatory compliance rather than reducing environmental operating costs, reducing impacts and liabilities and seeking commercial advantages.

Moreover it is not just within organisations that communication is important, traditional external mechanisms such as mandatory company reporting are relevant. For example, in the UK the President of the Institute of Chartered Accounts of England and Wales has recently indicated (FT, 1998) that annual company reports are failing to identify what risks are faced by a company, more explicitly, those dealing with the environment, society or ethics.

Any person concerned with environmental risks must therefore be cognisant of these developments in communication and integration along with their likely implications. Mutual understanding will be vital to avoid errors. One issue that will be key to broader integration is consistency in language. The term 'environment' is particularly vulnerable to misunderstanding. It can be viewed differently in the ecological/pollution context that is the primary focus here. However it has also become a popular term in business circles in its 'broader surroundings' meaning across a highlighted spectrum of applications. Waring and Glendon (1998) also note the lack of an agreement on the scope of risks and a common technical language as barriers to integration. Furthermore specialists in different risk-related disciplines are likely to value their separateness for fear that integration will lead to a loss of priority for their area.

Environmental risk terms may have explicit meanings unlike their 'everyday' use, for example some people view environmental risks as those aspects of the environment, such as storms or floods, that impact on their business rather than vice versa. The nature of the risks being discussed must be made explicit, along with details of risk receptors and the quantification techniques (if any) employed.

It is also important to differentiate between organisational and ecosystem risks. While there is clearly an overlap between the two, they are different in both their nature and the way in which they are measured.

The former is often assigned a monetary value or has a bearing on the overall public image or reputation of the organisation. Ecosystem risks by way of comparison would include potential damage to a watercourse or other natural resource or human health risks associated with an emission or discharge. This publication focuses principally on the former and the associated management tools that may be applied. This will of course facilitate the overriding objective of minimising ecosystem risks.

A UK government guide to risk assessment for environmental protection (DoE, 1995) gives the following risk-related definitions.

Hazard	A property or situation that in particular circumstances could lead to harm.
Risk	A combination of the probability, or frequency, of occurrence of a defined hazard and the magnitude of the consequences of the occurrence.
Risk assessment	Risk estimation and risk evaluation, where risk estimation is concerned with the outcome or consequences of an intention taking account of the probability of occurrence, the risk evaluation is concerned with determining the significance of the estimated risks for those affected.
Risk management	The process of implementing decisions about accepting or altering risks.

These complement earlier Royal Society Study Group Report recommendations (Royal Society, 1992).

Environmental hazard is an event or continuing process, which, if realised will lead to circumstances having the potential to degrade, directly or indirectly, the quality of the environment in the short or longer term.

Environmental risk is a measure of potential threats to the environment and combines the probability that events will cause or lead to degradation of the environment and the severity of that degradation. Liabilities such as financial obligations may come from the realisation of environmental risks.

These definitions provide a framework, however, problems arise because environmental risk assessment can be used as a term to describe techniques which, although superficially similar, measure very different things and give different results. Furthermore, the phrase 'we have done a risk assessment to support our action' should not be seen as a conclusion, but rather a logical framework facilitating questions (if necessary) on the validity of data, assumptions made and conclusions drawn.

Criticisms that have been levelled at risk-based approaches include the fact that they explicitly recognise that risks exist but that subsequently no resources might be allocated for their amelioration (the 'no action' scenario), or indeed, that the level of uncertainties involved in risk estimation is so great that the whole exercise cannot be justified. A response to these criticisms could be that it is not the aim of risk management to justify inaction, rather the aim is to maximise the benefit that can be achieved. It is therefore appropriate, for example, to focus on the assessment and remedial effort at a contaminated land site that poses an immediate threat to ecosystems or human health, rather than at a site which, despite known contamination, gives rise to a much lower level of risk. In addition, it is certainly the case that risk assessment highlights areas in which there are significant uncertainties; however, by recognising and making explicit these shortcomings, areas for further research are identified.

Finally, in all the preceding discussions it has been implicit that the risks assessed have been due to (principally) synthetic chemicals arising out of human activity. However, how should the risk associated with naturally occurring hazardous materials be addressed (eg natural levels of arsenic occurring in certain areas?) Moreover how would we deal with (as yet largely unknown) toxicity associated with chronic exposure to natural food constituents? Rodricks (1994) notes that in the case of the few natural chemicals in food that have been studied, a large fraction has proved capable of producing the same range and forms of toxicity that has been found for synthetic chemicals. He conjectures that if toxicologists were to subject the natural chemical constituents of the human diet to the same type of chronic toxicity testing now used for industrial chemicals, a large fraction will be found to produce serious forms of toxicity including cancer, birth defects and so forth.

Any attempt to broaden the scope in an already wide field must of course mean that significant amounts of material cannot be considered. There is an extensive literature on all of the aspects not addressed in this publication. The reader is directed to other articles or books where, in the opinion of the author, they have provided particular insights. An excellent starting point is the *Handbook of Environmental Risk Assessment and Management* (Calow, 1997) which devotes chapters to detailed discussions on issues such as the psychology of risk and communication, complementing coverage of more general issues. For an informed and challenging assessment of the important risk management issues facing organisations going forward into the twenty-first century the recent publication by Waring and Glendon (1998) is recommended. It is particularly strong on the identification of potential errors and inconsistencies in the application of risk management tools such as management system standards.

A proportion of material from the earlier publication has been reproduced here as it still holds a contemporary relevance. In particular, integration with health and safety and the frameworks such as those associated with the US Superfund programme. The title has been changed slightly to reflect the wider audience at which it is aimed. There are clearly dangers in attempting to appeal to a broad-cross functional audience, for example, a brief introduction to environmental management system standards is unlikely to convey any new insights to those with existing environmental responsibilities. Nonetheless it should hopefully demonstrate similarities in terms of risk identification and control mechanisms. Any feedback on the extent to which the publication fails or succeeds in its aims is most welcome.

CHAPTER 2
THE NATURE OF ENVIRONMENTAL RISKS

When considering what constitutes an environmental risk in terms of emissions from a site, it can be helpful to view it as consisting of three components – hazard, pathway and receptor. The magnitude of the hazard is related to the nature and quantity of materials and/or process that constitute the risk source. In itself the presence of a hazard might not be significant, as a pathway is required to mean that the hazard can be transported to a target. The nature of the receptor (or target) completes the overall risk profile. A combination of the three factors will determine how significant a risk exists by effectively considering what the probability is of the adverse event and what the consequences would be.

- *Hazards:* Typical hazards that might contribute to the environmental risk level at a site might include the storage of hazardous chemicals or fuels, the production of hazardous waste, the types of construction materials used in the building and ground contamination from current or former site use.

- *Pathways:* In order for hazards to affect potential receptors it is necessary for there to be some form of pathway. Where the hazard is liquid, typical pathways will be either the surface water drainage system, foul water drainage system or, for liquids stored in underground tanks, suitable geology. If the hazard is some form of particulate (eg asbestos dust) or excess noise, the pathway would be the atmosphere. Pathways for these potential hazards can be limited by suitable controls (eg interceptors for oils on the drainage system).

- *Receptors:* The final major element that affects the potential environmental risk from a site is the receptors. Even if there are a wide range of hazards associated with the site, the absence of sensitive receptors may give the site a lower overall risk. Typical receptors would include local watercourses, groundwater, soil and ecosystems, the atmosphere, neighbouring properties, employees or members of the public. The scale associated with the receptor can vary widely. There may be purely local concerns such as those involving neighbours in noise, visual intrusion or other nuisance issues. Acidic atmospheric emissions might cause impacts on a regional scale and greenhouse gases would give rise to global concerns.

A variety of controls can be utilised at a site to reduce the potential hazards outlined above. Such controls might include bunding of hazardous fuels or chemical tanks, ensuring that hazardous chemicals or waste are disposed of in an appropriate manner or implementing an environmental management system. A wide range of controls could be utilised at a site depending on site-specific conditions. The introduction of controls will tend to reduce the potential for hazards to pose a risk to receptors, and therefore decrease the environmental risk at a site.

In terms of materials that represent environmental hazards, the list extends beyond the agents such as petrochemicals, pesticides and chlorinated solvents that are normally considered. Natural products such as milk or fruit juice can cause damage, particularly if released to water. In a UK incident, apple juice was washed out of tanks via storm drains and into a nearby watercourse, The juice depleted dissolved oxygen in the river and caused fish deaths as well as damaging a local water treatment works. The company responsible received an £11,000 fine and was ordered to pay £19,000 costs. This incident also highlights the fact that, in many instances, the most significant financial liabilities arise not from fines for breaches of legislation but for the associated clean-up works.

A risk-based approach finds no difficulty in dealing with emissions of potentially hazardous substances that are legally sanctioned, for example, permitted in the discharge consent. A risk management approach would not simply accept compliance as a satisfactory position, but would rather view the emission as an ongoing risk that could potentially be reduced. This is a more realistic stance in many ways as it is quite possible that legal standards will change and require action sometime in the future. A risk-based approach (rather than compliance) can therefore place greater emphasis on performance improvement, consistent with environmental management standards such as ISO 14001.

Although one of the principal ideas underpinning this publication is that legal compliance should not be regarded as the ultimate aim in environmental management, it is also evident that legislation forms a very important baseline. It also serves as a measure of more general concern. The recent fine imposed on the Milford Haven Port Authority under the Water Resources Act of £4M relating to the *Sea Empress* oil spill highlights this, being the largest ever fine awarded in a pollution case and as such might be considered a landmark ruling (Environment Business, 1999a). The judge also noted that he considered the fine to be considerably lower than would be appropriate for a major company with significant resources.

There are many areas of potential uncertainties associated with a 'conventional' human health-based risk assessment. For example, toxicity information derived from animal studies is extrapolated to apply to consider effects on humans. Nevertheless, the difficulties are multiplied when environmental risk assessments are undertaken. In human health terms, there may be more sensitive populations exposed such as children or elderly people. The exposure patterns may also be more difficult to estimate (and control) and might not be restricted to work hours, for example, if there are houses adjacent to a hazard source.

In relation to risk to humans it is clear that the life and well being of a single individual is very highly valued. In the case of other forms of life it may be that the principal concern is for the species or community rather than the individual. That is not to say that the deaths of individual animals is not of concern. Clearly in cases of, for example, the killing of fish, there is legal protection that might result in prosecution for the offending party. The decision on whether an individual or a species is of principal concern therefore needs to be considered. An exposure to a substance may harm an individual earthworm, but the overall population may adapt and be relatively unaffected, for example. The wide range of substances that could interact and lead to additive or synergistic effects is also relevant.

Extending the scope of an environmental risk assessment to consider ecosystems is also likely to give rise to difficulties in identifying appropriate toxicity information for relevant flora and fauna. The most important extra considerations are, however, interaction with other environmental components (eg adsorption onto sediments and degradation reactions that determine the persistence and interaction with living organisms could lead to bioaccumulation). Physiochemical properties of substances such as water solubility and volatility are therefore important in determining the risk associated with substances. However, two biologically mediated mechanisms – biodegradation and bioaccumulation – are of particular interest.

Although biodegradability is usually seen as a 'good thing', the environmental impacts of some environmental incidents (eg silage or food products released into water) arise out of this property. As the substance breaks down, it uses up dissolved oxygen in the water. This then leads to fish kill and other damage. Biochemical Oxygen Demand (BOD) and Chemical Oxygen Demand (COD) are often used as indicators of degradability. They represent the amount of dissolved oxygen used up during biologically mediated breakdown and following reaction with a strong chemical oxidising agent respectively. Degradation is often quoted in terms of a half life ($t_{}$) which is the time taken for half of the substance present to degrade.

Bioaccumulation occurs where some substances transfer preferentially into living organisms from aqueous media, because of an affinity, for example, with fatty tissues (lipophilism). This process can result in the accumulation of substances to dangerous levels within organisms or even biomagnification up the food chain. Problems of DDT being biomagnified up the food chain to affect bird of prey reproduction is probably the best known example.

Sources of Environmental Information

A good initial source of environmental risk information on a product or substance is the information required for the supply or transport of the agent (which primarily relates to health and safety concerns but is being continually extended to include specific environmental information). These aspects are important because they establish the benchmark for the type of environmental information that might be obtained from suppliers.

The EC Dangerous Substances Directive and Dangerous Preparations Directive, 91/155/EEC (OJL 76-22-3-91) and 93/112/EEC (OJL 314 16-12-93), indicates that safety data sheets should contain the following information:

- identification of the substance/preparation and the company supplying it;
- composition/information on ingredients;
- information on hazards;
- first-aid measures;
- fire-fighting measures;
- accidental release measures;
- handling and storage;
- exposure controls and personal protection;
- physico-chemical properties;
- stability and reactivity;
- toxicological information;
- disposal considerations;
- transport information;
- regulatory information; and
- other information.

In the UK, the HSE approved code of practice (ACOP) on safety data sheets (HSE, 1995), gives examples of relevant ecological information for inclusion in a safety data sheet, including

a) Mobility
   i) known or predicted distribution to environmental compartments
   ii) surface tension
   iii) absorption/desorption
   iv) other relevant physical properties

b) Degradability
   i) biotic and abiotic degradation
   ii) aerobic and anaerobic degradation
   iii) persistence

c) Accumulation
   i) bioaccumulation potential
   ii) biomagnification

d) Short and long-term effects on
   i) Ecotoxicity
      - aquatic organisms
      - soil organisms
      - plants and terrestrial animals
   ii) Other adverse effects
      - ozone depletion potential
      - photochemical ozone creation
      - global warming potential
      - effects on wastewater treatment plants

The ACOP indicated that information relevant to the environment should also be included in other sections of the safety data sheet, especially advice for controlled release, accidental release measures and disposal considerations. This should include appropriate methods of disposal, (eg incineration, landfill, recycling etc). In cases where the evaluation of preparations has not been undertaken, account should be made of substances present in the preparation.

Out of the Dangerous Substances Directive (92/93/EEC (OJL 154 5.6. 92) and 93/21/EEC (OJL 110/110A 4.5 93)), the UK CHIP Regulations require:

a) the identification of the dangerous properties of the substance through the evaluation of available data on environmental impacts

b) environmental labelling of substances – substances classified as dangerous for the environment have to bear the appropriate hazard symbol (dead fish and a leafless tree).

A number of environmental risk phrases apply:

- Aquatic Environment

  R50 Very toxic to aquatic organisms

  R51 Toxic to aquatic organisms

R52 Harmful to aquatic organisms
R53 May cause long-term adverse effects in the aquatic environment

- Non-aquatic Environment
  R54 Toxic to flora
  R55 Toxic to fauna
  R56 Toxic to soil organisms
  R57 Toxic to bees
  R58 May cause long-term adverse effects in the environment (other than the aquatic environment)
  R59 Dangerous for the ozone layer

The R59 phrase re-emphasises the point made earlier that some environmentally hazardous substances such as the stratospheric ozone depleting halons and chlorfluorocarbons (CFCs) have not had major health and safety concerns, but pose environmental threats.

Release Scenarios

In considering environmental risks it is appropriate to assess what scenarios might be relevant for the release of environmentally hazardous materials. A review (Pearson, 1988) has described potential release scenarios associated with storage of hazardous materials in a manner which is of broader relevance. Storage facilities such as warehouses are common to virtually all industrial activities and the potential risks associated with these hazardous materials are therefore of concern to virtually all businesses. In addition, warehouses are often situated near major waterways, considered to be the environmental phase most at risk from releases.

- Major release of gas or vapour
  Gas or vapour clouds will tend to disperse rapidly, and it was considered unlikely that any gas release will have a greater impact on wildlife than on man.

- Release of non-volatile liquid
  Opening discharge valves by accident or design is a well-documented event.

- Fire or explosion leading to direct release to air or water
  Shock waves from explosions or the effect of fire causing the direct rupture of tanks or the collapse of racking indirectly leading to rupture.

- Combustion products
  Combustion products may be produced in the course of a fire that can lead to local concern; dusts and other materials may be transported and deposited in the site vicinity. In some instances toxic materials may arise from the incomplete combustion of precursors such as polymers which are viewed as harmless.

- Fire-fighting water
  The large volumes of water used in fire fighting can carry liquid and solids (eg from ruptured containers) into the site drainage system and then into water bodies.

- Vehicle movement
  Vehicles have significant potential to cause damage to racks, tanks and packaging. Attention should therefore be given to delivery routes in relation both to underground services such as drains and above ground items such as fuel transfer pipes.

- Site security
  Regular surveillance of unmanned sites is desirable to minimise the possibility of malicious opening of valves or starting of fires.

The review also suggested ways of reducing risk. For example, fire spread could be reduced by the use of fire resistant materials, fire check walls, water curtains and gaps between storage bays. Waterproofing of floors and bunding of storage areas would prevent uncontrolled leakages to ground and waters.

Maintenance of an up-to-date inventory of materials held in storage would enable decisions to be made on appropriate fire fighting. It was also said that provision could be made for the collection of firewater (for later treatment) where hazardous materials are stored. It was suggested that the commonly accepted figure of 2-5m^3 firewater per tonne of stored product might be inadequate, with up to 10m^3 per tonne being required in practice.

Environmental Pathways and Receptors

Two important factors must be considered when considering environmental risks, firstly environmental risks are strongly location dependent. The same operation in a different geographical location can represent very different risks. The second consideration is that zero-risk options are not usually available. In relation to the contamination of land, for example, both a no-action scenario (ie leave the contamination where it is) and the complete off-site removal of contaminants involves risks (which may be quite different in nature and difficult to compare). Removing soil contamination and depositing in a landfill site does not destroy the hazard, however the anticipated overall risk is lower if it is held in a waste facility designed for that purpose.

The site dependency of environmental risks is a key factor that differentiates them from health and safety concerns. It may be relatively straightforward to establish a hazard source – pathway – receptor link to assess risks for health and safety (usually depending on operators work behaviour) however, off-site information must be gathered to identify the pathway and receptor elements of environmental risk.

Information that is likely to be required on the pathway and target elements for water will therefore include local ground cover and soil type. The routes and destinations of both foul and surface water drains should also be determined. In terms of off site targets, elements such as fisheries and public water supplies are important for surface waters. The type and importance of local aquifers is also relevant since they may be used directly as a drinking water source or they may replenish surface water bodies.

A first step in this process would involve a review of maps, literature and other information, identifying off site environmental features of the area, including:

- population distribution and local residential patterns;

- flora and fauna with the identification of any important, protected or endangered species and areas of nature conservation values;

- architectural and historic heritage, including archaeological sites, ancient monuments and listed buildings;

- landscape and topographical features; and

- recreational areas, including public rights of way, footpaths and bridleways.

One of the key aspects to consider is the pollution risk to water. The classification of local aquifers and whether there are nearby abstractions as well as information on surface water quality is important.

Reference can be made to Department of the Environment, Transport and the Regions guidance (DETR, 1999) in determining what might constitute receptors or targets of releases. Its purpose is to provide help in industry and regulators in addressing environmental issues under the COMAH regulations (superseding earlier Department of the Environment guidance (DoE, 1991) on the CIMAH regulations). Although the guidance is aimed at specific, generally high hazard, sectors of industry it has a broader applicability. In identifying major accident hazards to the environment it highlights events which have the potential to:

- pose knock-on threats to human health by contamination of food and drinking water or impacting sewage treatment systems;
- affect large areas of land designated for conservation, amenity or planning purposes;
- inhibit natural regenerative processes; and
- cause significant ecosystem damage, such as reducing breeding success of protected species, damaging rare habitat or reducing biodiversity.

Criteria and thresholds are then considered individually for targets such as SSSIs. Emergency plans are also discussed along with provision for restoration and clean-up following a major accident.

Table 2.1 *Environmental Accidents*

Type of accident	*No of accidents with observed effects*
Ship collisions	22
Unintentional discharges from industrial tanks or process equipment	17
Fire or explosion in storage	5
Leaks on pipelines	3
Train derailment	2

Environmental phase affected	*No of accidents*
Rivers and lakes	31
Marine ecosystems	16
Terrestrial and rivers	4
Terrestrial	3
Groundwater	1

Much of the above discussion has been focused towards potential damage to ground or surface waters, a move that appears to be justified by a review that examined the literature on various accidents and incidents with ecological consequences over the period 1975 – 1990 (Lindgaard-Jorgensen and Bender, 1992). More than 1000 references to accidents were assessed of which approximately 50 per cent specifically included the term 'environment'. The types of accidents with observed ecological consequences are shown in Table 2.1.

It was noted that although only a small number of accidents contained information on observed ecological effects, the actual number might well have been much higher because in many instances accidents were not reported or causes were not explained.

CHAPTER 3
PROPERTY RELATED ENVIRONMENTAL RISKS

This chapter considers the identification and management of environmental risks associated with property. Every commercial operation with a physical base should therefore find this a significant issue. It is also an area where historical activities, completely unrelated to current or future operations, may have an impact – principally through the contamination of land. Furthermore, it can be relevant beyond the occupiers of the property through to managing agents, landowners, and property investors and on to financial institutions whose collateral on lent money is the value of the property.

Professional bodies such as the Royal Institution of Chartered Surveyors (RICS) have recognised the importance of environmental issues and considerable reference is made here to relevant publications and guidance. A tiered approach is presented that focuses on obtaining adequate information on which to base management decisions. Separate discussion is also presented on two property related topics, (contaminated land and fire) reflecting their particular significance. Contaminated land remains the principal property related concern, reflected by the current changes to the UK regulatory framework. It is an issue that can be relevant to commercial organisations that consider themselves in low hazard sectors, since they may acquire property with undetected contamination. Fire risks are rightly seen as a major business concern, however, the environmental aspects of these incidents usually receive less attention. They are discussed here with a view to ensuring that environmental considerations are accommodated within the broader risk management context.

The US experience with increasing regulatory requirements through the 1970s and 1980s, in particular Superfund, raised concern over the inheritance of other people's liabilities. This led to the employment of specialists to undertake environmental work as part of the due diligence process when acquiring property. The management approach to property risks has therefore been influenced by these property transfer protocols. Confusingly this type of work has been given a range of different names, including environmental audit or review, property transfer study, environmental site assessment or most commonly pre-acquisition audit (although parties other than a prospective purchaser may commission them).

Two factors – time and cost – were the principal drivers behind the adoption of a phased approach to this work. Detailed investigations of ground conditions can be both expensive and time consuming, usually requiring several weeks to complete. In the context of fast moving corporate transactions this length of delay can be detrimental. Similarly, a large proportion of sites can often be excluded from detailed investigation by simple screening to demonstrate that there are unlikely to be significant liabilities at the properties.

It should also be noted that risks are not just restricted to heavy industrial type operations. Turner et al (1994) noted that a broad range of activities have the potential to cause pollution, including light industry and the service sector. Furthermore, they suggest that a prudent property investor may need to look beyond the site-based operations to consider the 'indirect risk' associated with tenants operations at other non-owned properties. Contaminated land has been, and is likely to remain the most significant single aspect and its characteristics lend themselves to management by a risk-based approach. Table 3.1 gives an indication (not exclusive, however) of the broad range of activities that could give rise to contamination of land.

Table 3.1 *Uses of Land Which May Give Rise to Contamination*

Agriculture	Burial of diseased livestock.
Extractive industry	Extracting, handling and storage of carbonaceous materials such as coal, lignite, petroleum, natural gas, or bituminous shale.
Energy industry	Producing gas from coal, lignite, oil or other carbonaceous material (other than from sewage or other waste) or from mixtures of those materials.
Production of metals	Production, refining or recovery of metals by physical, chemical, thermal or electrolytic or other extraction processes.
Production of non-metals and their products	Production or refining of non-metals by treatment of the ore.
Glass making and ceramics	Manufacture of glass and products based on glass. Manufacture of ceramics and products based on ceramics, including glazes and vitreous enamel.
Production and use of chemicals	Production, refining, recovery or storage of petroleum or petrochemicals or their by-products, including tar and bitumen processes and the manufacture of asphalt.
Engineering and manufacturing processes	Manufacture of metal goods, including mechanical engineering, industrial plant or steelwork, motor vehicles, ships, railways or tramway vehicles, aircraft, aerospace equipment or similar equipment. Storage, manufacture or testing of explosives, propellants, ordnance, small arms or ammunition. Manufacture and repair of electrical and electronic components and equipment.
Food processing industry	Manufacture of pet foods or animal feedstuffs. Processing of animal by-products (including rendering and maggot farming, but excluding slaughterhouses, butchering).
Paper, pulp and printing industry	Making of paper pulp, paper or board, or paper or board products, including printing and de-inking.
Timber and timber products	Chemical treatment and coating of timber and timber products.
Textile industry	Tanning, dressing, fellmongering or other processes for preparing, treating or working leather. Fulling, bleaching, dyeing or other textile floor coverings (including linoleum works).
Rubber industry	Processing of natural or synthetic rubber (including tyre manufacture or retreading).
Infrastructure	Marshalling, dismantling, repairing or maintenance of railway rolling stock. Dismantling, repairing or maintenance of marine vessels, including hovercraft. Dismantling, repairing or maintenance of road transport or road haulage vehicles. Dismantling, repairing or maintenance of air or space transport systems.
Water disposal Miscellaneous	Treating of sewage or other effluent. Storage, treatment or disposal of sludge including sludge from water treatment works. Treating, keeping, depositing or disposing of waste, including scrap (to include infilled canal basins, docks or rivercourse). Storage or disposal of radioactive materials Premises housing dry cleaning operations. Laboratories for educational or research purposes. Demolition of buildings, plant or equipment used for any of the activities in the schedule.

Source: Contaminated Land and the Water Environment, Report of the National Rivers Authority, Water Quality Series, No 15, March 1994

Some of the factors that contribute to the overall level of environmental risks associated with a property are shown in Table 3.2. Clearly, one of the most significant elements is the nature of site activity, with industrial-type operations likely to be significantly higher risk than service organisations. The age of the site is important for several reasons. For example, environmental control standards such as containment around bulk liquid storage tanks are generally higher for more recent installations. There is also the use of hazardous materials such as asbestos, which are more likely to be found in a building of 1960s or 1970s vintage than a recent building. The production of hazardous waste can also be a significant factor. Industrial-type operators are more likely to produce hazardous solid or liquid wastes; but service operations can also produce environmentally hazardous wastes, for example, restaurants, pharmacists and photographic shops. The proximity of nearby watercourses or vulnerable and sensitive hydrogeology is a significant consideration as these receptors are perhaps those most liable to suffer in an environmental incident.

Table 3.2 *Environmental Risk Factors*

Risk increasing	*Risk reducing*
Older Site	Younger Site
Process Industries	Service Industries
Owners/tenants without environmental management controls	Owners/tenants with environmental management controls
Hazardous waste produced	Low waste production, particularly if non-hazardous waste produced
Lack of knowledge of drainage system	Well understood and controlled drainage system
Low standards of bulk liquid containment	High standards of bulk liquid containment
Evidence of land contamination or contaminative uses	No evidence of contamination or contaminative uses
Sensitive or potentially polluting neighbours	Non-sensitive or benign neighbours
Proximity of and discharges to watercourses	Absence of sensitive neighbouring watercourses
Poor housekeeping	Good housekeeping
Complaints, warnings or prosecutions	No compliance or nuisance issues
Risk-increasing geology and hydrogeology	Impermeable soils and no sensitive aquifers

An important aspect to consider in risk management is that of risk perception, linked in a property context to stigma. It is accepted that risks are likely to be perceived differently by different groups of individuals, which can have an important influence on government (and organisation) policy. Syms (1996) explored this issue in the context of property professionals. Property valuers' perception of environmental risks was found to align closely with that of the general population. These perceptions, however, differed from the 'non-valuer' property professional whose ranks included engineers, lawyers and environmental scientists. It was postulated that this result reflects the greater understanding of environmental risk by the non-valuers. This might be important as valuers may well adopt a very cautious approach and overcompensate for their incomplete knowledge.

A report (Fink, 1998) on a recent Court of Appeal finding is pertinent to the stigma and risk perception debate. The disclosure of radioactive contamination during the sale of a site adjacent to the Aldermaston Atomic Weapons Establishment meant that the deal collapsed. An action for damages was then brought against the Ministry of Defence, and the court awarded three quarters of the purchase cost (less the diminished value of the land) and running costs. On appeal the issue hinged on whether compensation was appropriate for only that part of the property affected (an area of marshland) or the diminution in the value of the entire property (including that attributed to stigmatisation). The ruling was in favour of the entire property value. There are some novel features in this case such as the claim being brought under the Nuclear Installations Act, nevertheless it is significant in terms of how stigma (or perceived risk) is addressed.

Notwithstanding the items referenced above, there has been relatively little systematic study on the impact of environmental risk (in the form of land contamination) on property values. A 1997 report by Richards for the College of Estate Management is therefore particularly pertinent. This thoughtful research observed that allowances were being routinely made (in 1996) for the impacts of contamination. However, these were invariably being made on the basis of maximum potential cost rather than 'most likely' or probabilistic approaches. It was also found that in over one third of cases where purchasers walked away from property investments affected by contamination, they had not conducted any form of environmental assessment. The approaches to valuation were found to be most commonly based on a cost to correct approach (ie deducting remediation costs from an unimpaired, clean value and incorporating an allowance for stigma). Within such approaches it was found that a single remediation cost figure was suggested as being incorporated within the calculation. Richards observed that valuations would be more realistic if some contingency were made in the allowance for remediation costs or a more sophisticated probabilistic approach was employed, possibly using Monte Carlo techniques.

The uncertainty regarding financial aspects is reflected in the historical difficulties in obtaining insurance cover for property environmental risks. Nonetheless, there appears to be some progress in this area. The likely growth in environmental insurance products available to property owners and investors was discussed in a recent article (Hill, 1999) along with the likely dual role of insurers. In addition to providing cover, the author noted that insurers and brokers are likely to be working with clients on environmental risk assessment and management protocols.

Property Risk Management

The management of property-related environmental risks presents some particular challenges. The cost of obtaining information on which to base decision-making can be expensive and involve significant use of external specialists. In addition, the timescales involved in assessment can cause problems, for example, contaminated land and its remediation might require periods of months, which can jeopardise corporate transactions and lead to increased cost uncertainty.

There are likely to be a range of existing risk management mechanisms in place within an organisation, such as those linked to property survey and valuations. General site information obtained for non-environmental reasons ranging from basic aspects such as location and activities through to housekeeping and health and safety, form a useful starting point. It is clear therefore that no single approach will be appropriate for all organisations, and the approach described here should not be regarded as prescriptive. Every organisation should start with a review of existing controls and assess the degree to which they meet environmental management requirements.

A tiered approach to the collection of information links the degree (and cost) of investigation to the amount of information that is required to make a management decision. Possible elements in a property portfolio risk management process are highlighted in Figure 3.1. It shows that a great deal of information can be obtained without invoking the Phase 1-2-3 studies normally associated with sale or purchase. It is likely that a formal Phase 1 study would be conducted in case of purchase, but it may not be necessary for the ongoing management of currently held property. Existing information and survey processes may also be utilised with supplementary elements to provide sufficient information for the ongoing management of environmental concerns. For example, Barclays Environmental Risk Management Unit employ a stepped approach starting from a simple Land Use Questionnaire (LUQ) used by valuers leading on to a more detailed questionnaire before a simple desk-based assessment and a Phase 1 study is commissioned. A 1998 study across their UK branches showed that no further action was taken in 89 per cent of cases when an LUQ was completed. A more detailed environmental questionnaire was required in 9 per cent of cases. In only 3 per cent of cases was further work such as a Phase 1 study required (Wright, 1998).

The Property Observation Checklist (reproduced here as Figure 3.2 with the permission of RICS) forms an excellent starting point and may well provide sufficient information for the management of sites that have been owned for some time. However, should particular concerns be raised about a site or an acquisition or divestment be undertaken, it is likely that further data collection is desirable. This could produce a site risk profile.

Typical management actions arising out of site risk profiles could include immediate closer inspection, changing the survey period, considering the relocation of hazardous activities or opening a dialogue with the regulator or other local stakeholders. Of course the availability of site information should minimise the likelihood of unpleasant surprises during any sale or purchase when this eventuality arises.

There are three principal means of obtaining information that identify a site risk profile:

1	direct observation	eg visual evidence of contamination such as ground-staining
2	interview	eg asking site staff for information on historic operations
3	desk/office study	eg identifying groundwater vulnerability in the vicinity of a site

There is obviously some overlap between these approaches, for example, site personnel may have information on previous industrial practices or any incidents that may have occurred. Indeed, it may often be useful trying to obtain the same information from more than one source, firstly as a confirmation mechanism, but more importantly because environmental data sources are often incomplete (a particular example here would be records of waste disposal) or even inaccurate. A major issue arising could be the control of site risk due to current operations. In such cases attention is likely to focus on control measures such as those covered by ISO 14001.

Details of current environmental management controls will give an indication of the ongoing control of site hazards. Supplementary information might be obtained from third party audits or reviews in addition to interview and inspection. This information could be brought together in an environmental management profile.

Notwithstanding the above discussion, a significant proportion of this chapter addresses the property transfer issue identified in Figure 3.1 as leading on from site risk profiles. This reflects not only the importance of this element but also the fact that it is an area that has seen the development of recognised standards. A Phase 1 study is often termed a pre-acquisition audit or property transfer assessment, indicating one of its primary applications. It is also being studied by the International Standards Organisation (ISO) who are assessing the possibility of producing a global standard on the topic. More detailed, expensive and time consuming studies such as the Phase 2 (usually involving ground investigation and laboratory analysis) and Phase 3 (involving detailed ground remediation assessment) are almost always undertaken by external specialists.

Although discrete items are identified in Figure 3.1 it is likely that each individual organisation would wish to utilise internal systems in unique ways that would not fit neatly into the boxes in the figure. The benefits of a tiered approach (as opposed to moving directly to a more detailed study) are not necessarily restricted to more effective use of finances. Undertaking preliminary work in-house will allow staff to gain greater understanding of the fundamental issues and facilitate management decisions.

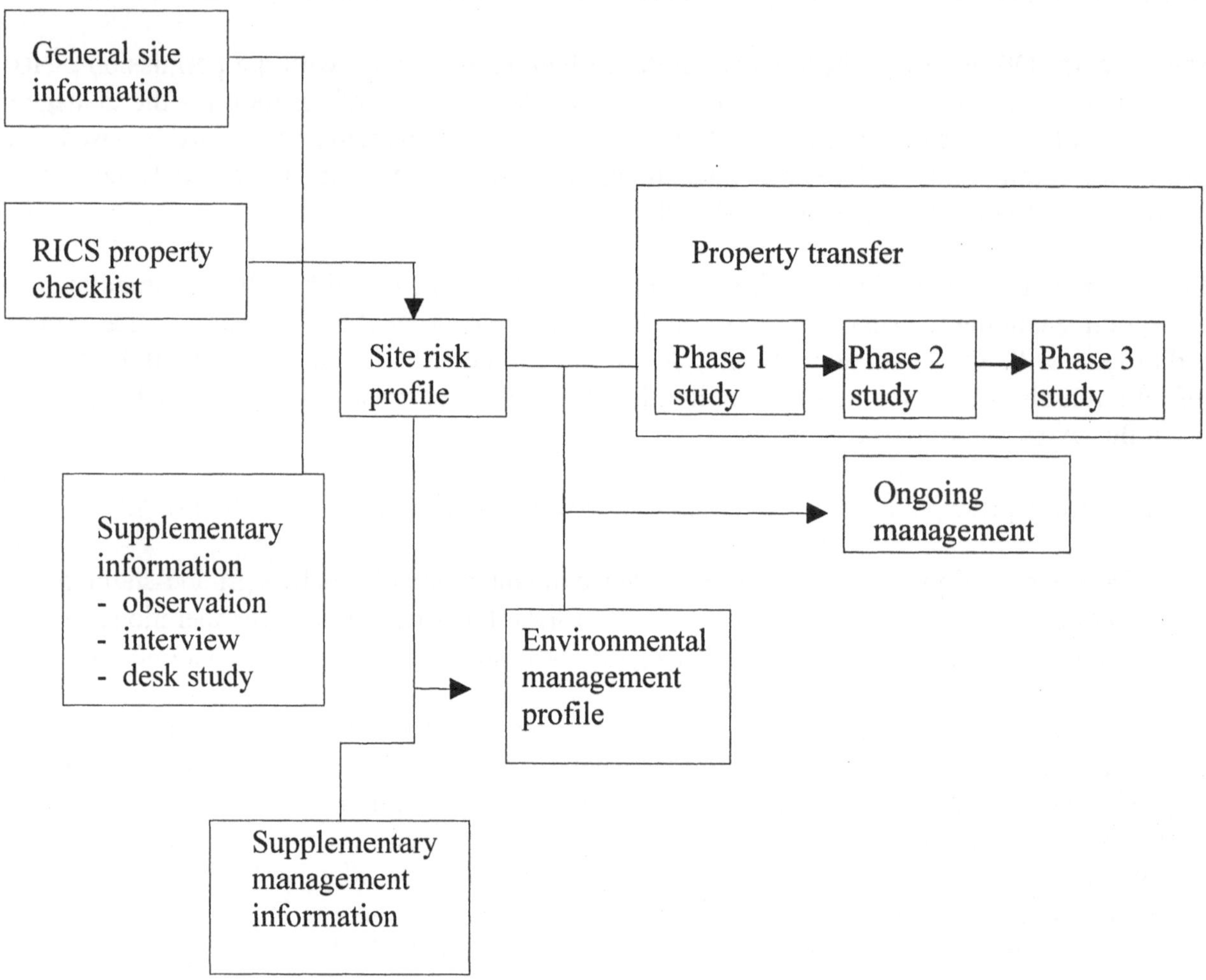

Figure 3.1 *Property Risk Management Process*

APPENDIX B

PROPERTY OBSERVATION CHECKLIST FOR IDENTIFYING APPARENT POTENTIAL FOR CONTAMINATION

Completion of this checklist is not to be construed as an environmental audit

Site/Property: Date of Inspection: By:

Tick ☑ the appropriate boxes to record presence and inspection of the various parts of the subject property and features observed in the course of the ordinary inspection of the property and the surrounding area.

PART OF PROPERTY							CONTAMINATION: FEATURES									
			Present Y	Present N	Inspected Y	Inspected N	Asbestos	Waste/fly tipping	Petroleum or fuel tanks	Other tanks or containers	Chemical odours	Discoloured or smelly water/ liquids	Oil staining	Irregular topography	Vegetation die-back	High voltage overhead tension lines or electrical sub-stations
Buildings:	Y	N														
occupied?																
vacant?																
dilapidated?																
Basements and subterranean features (e.g. air raid shelters)																
Outbuildings, including garages and stores																
Hardstanding																
Agricultural land																
Derelict or rough ground																
Landscaping and/or natural woodland																
Watercourse or adjacent thereto																
Surrounding area																
Extent of inspection of immediate vicinity																

Note: This checklist is not designed for use in respect of residential property or agricultural land.

List existing uses:

	Observed use of the subject property:		Evidence observed of such uses within the vicinity:	
Industry/manufacturing?	☐ Yes	☐ No	☐ Yes	☐ No
Waste disposal, waste processing and/or landfill?	☐ Yes	☐ No	☐ Yes	☐ No
Surface or underground mineral working?	☐ Yes	☐ No	☐ Yes	☐ No
Vehicle maintenance or refuelling?	☐ Yes	☐ No	☐ Yes	☐ No
Was any watercourse observed within the vicinity?			☐ Yes	☐ No

NOTES OR OTHER COMMENTS

Figure 3.2 *RICS Property Observation Checklist*

RICS Property Observation Checklist

The Royal Institution of Chartered Surveyors has recognised that their members can be involved in decisions relating to environmental risk, particularly land contamination (RICS, 1997). A guidance note has been produced to assist chartered surveyors who are not environmental specialists to define their responsibilities and to provide assistance on the identification and investigation of contaminated land.

They note that, in the absence of instructions or agreement to the contrary, completion of the property observation checklist in the guidance note is regarded, in relation to a valuer's inspection responsibilities, as consistent with duties under Practice Statements PS 6.1 and PS 6.3.1 of its *Appraisal and Valuation Manual.* Should surveyors be used to obtain additional site information it would be essential that an agreement be reached (preferably in writing) on the basis of the surveyor checking that the additional information required is within:

1 their competence;

2 the scope of their PI insurance.

On this second point it is likely that clarification will be required from their insurers. It should be emphasised nevertheless that the additional site information (such as that described in the following section) required to give a site risk profile is relatively straightforward and complements the RICS checklist reproduced here as Figure 3.2. The RICS state that although few surveyors have the specialist knowledge or expertise to investigate contamination '*whatever the agreement surveyors have a duty to report possible contamination seen in the course of their inspections*'. The surveyor using the checklist is expected to tick the boxes relating to the particular feature and make additional notes as necessary.

The majority of questions relate to directly observable contamination on the site such as ground staining, however, the presence of fuel tanks and vehicle maintenance suggests potential for (rather than actual) ground contamination. Questions are also asked about neighbouring activities (which is important as contamination might migrate from their property onto the site). Importantly, the presence of watercourses is highlighted. This is very important because it is one of the key targets for contamination, hence it raises the overall site sensitivity. Accordingly, any contamination present on a site may well assume a greater significance in the proximity of a river into which pollutants can migrate. In addition, should an accident occur in the future, the potential consequences of a release could be much greater.

Property Condition Checklist – Supporting Information

The RICS checklist represents a useful model for site inspection that is supplemented here by a number of items arising out of observations from site audits. These specifically refer to UK sites and derive directly from the author's experience.

- The nature of adjoining properties is relevant to the assessment of potential liabilities. For example a noisy process in an urban area might be regarded as acceptable but could be a potential nuisance in a rural setting. There may be other factors near to the site that make it environmentally sensitive, such as the presence of protected animals like badgers and bats, or nature reserves and Sites of Special Scientific Interest.

- Potential off-site sources of contamination could include, for example, nearby petrol station activities that might lead to contamination of the soil and water that could then migrate onto the surveyed property.

- Indicators of possible ground contamination on a site could include: stained soil, seeps/leachate, stressed or absent vegetation, odours, unusual surface water characteristics, dead fish, discoloration including any oil sheens, foaming etc, unusual topographic features, areas of subsidence and excavated/filled areas.

- Use or storage on the site of organic solvents, petrol, diesel, fuel oil etc could act as an indicator of potential liabilities due to leaks and/or spills. Above ground tanks should normally be inside a containment wall (bund) capable of retaining the contents of the tank in case of a leak and should therefore retain, for example, rainwater that will need to be emptied periodically by site staff.

- Underground storage tanks and associated pipework are of particular concern as they may have leaks that go undetected over a period of many years. Generally, the older the tank the more likely it is to be a 'leaker'. Petrol tanks are required to be registered and leak-tested periodically. However, other storage tanks (eg for oil) are not subject to this requirement. Should tanks have been taken out of commission it should be checked whether they still contain liquids.

- The concern over a water abstraction in or near the site would be that of contamination of the supply. Should this be suspected a request could be made for details of any chemical analyses undertaken on the supply.

- Drains are very important pathways, indeed one of the commonest ways rivers are affected is by spills or leaks escaping into surface water drains. It should be checked whether there have been problems with the site drainage, such as cracked or collapsed drains, backing up or blocking of drains, unusual odours or staining. Efforts should be made in the site visit to inspect the site drainage, particularly at discharge points to rivers etc, for evidence of contamination by materials such as oil.

- Should the site have an interceptor (sometimes called a separator) installed in the drainage system it may well be that there are (or have been) underground storage tanks or some other potential source of contamination present on the site.

- Industrial processes may produce effluents that might be acidic or alkaline or contain chemical contaminants such as metals. In these instances a discharge consent, with certain conditions attached, would be required from the local water company to discharge to sewers. Effluent from WCs and showers does not require a consent, however, vehicle washes are usually subject to a consent. Sometimes a wastewater treatment plant exists on site to remove contaminants before discharge to sewers.

- Information on waste management, both current and historical, is very important in assessing potential liabilities. The nature of wastes produced by the site and its neighbours and whether it was removed off-site for disposal are important. Burning or burying waste on the site should be identified where practicable. Illegal dumping (flytipping) on properties can also cause problems.

- Most wastes are subject to the Duty of Care Regulations that require that wastes are stored, packaged and transported safely. The site management should therefore be aware of the types of waste they are producing in order that they might be managed properly. There are more stringent legal requirements applying to hazardous wastes in the UK (these are known as 'Special Wastes') such as asbestos, acidic, flammable or toxic materials.

- Significant atmospheric discharges may require prior authorisation by either the local authority or by the Environment Agency. These are generally industrial process type operations. Smells, dusts or smoke may cause nuisance to the neighbouring community and, as such, render the operator liable to prosecution.

- Equipment within electricity substations often contains large quantities of oil. Historically this may have been contaminated with toxic PCBs (polychlorinated bipenyls). Often the substations belong to the local Electricity Company who could provide information on the oil. Should the transformers be relatively old and owned by the site, some effort might be necessary to verify that they are PCB free (normally a chemical screening test).

Site Risk Profiles

When assessing the overall level of environmental risk that a property or site can pose to the environment it is important to recognise that it will depend on a combination of several factors as discussed in Chapter 2, namely:

- Hazards
- Controls
- Pathways
- Receptors

When the site risk profile is drawn up the constituent elements must therefore be considered. In the environmental risk assessment for each site the combined effect of these factors should be assessed to obtain the risk ranking. Many sites have few hazards associated with them. Hence, even if there are few controls, obvious pathways and sensitive receptors, they are not likely to have a high risk ranking. Other sites that do have significant hazards associated with them may have minimal controls as well as potential pathways and receptors and hence can pose a greater environmental risk.

Five risk categories that could be assigned to sites are:

1 Potentially Significant
2 Higher than Average
3 Average
4 Lower than Average
5 Likely to be Insignificant

The categorisation is a relative measure, avoiding the use of terms such as high-risk and low-risk. The recommendations below are of a general nature and are based on the risk category to which a site has been assigned. For sites in each risk category the following example recommendations could be made:

- *Potentially Significant: A visit by an environmental specialist is recommended for these sites. The report produced would consider risk associated with current operations complemented by a desk study that would include consideration of historical contaminative uses.*

- *Higher than Average: A visit within the longer term by an environmental specialist is recommended. The scope of these visits would be similar to that described for the potentially significant sites.*

- *Average: Sites assigned to this category are put on a one year review cycle, although resolution of site specific recommendations might be achieved before that time. The routine property surveyors' reports form the principal basis of the review process.*

- *Lower than Average: These sites are placed on a two-year review cycle.*

- *Likely to be Insignificant: These properties are placed on a three-year review cycle.*

Naturally there may be a need to revise these categories in the light of experience. For example, should incidents be reported at a lower risk site it would be worthwhile to reassign it to a higher risk category.

Supplementary Information and Site Risk Profiles

The completed Property Condition Environmental Checklist and complementary information can be assessed to enable a risk categorisation. Information on the groundwater and surface water vulnerability of the site may be taken from reference sources and included in the assessment. A brief report can then be produced indicating key risk-related points, categorisation and any arising site-specific recommendations. It is important to recognise that there is a subjective element in assessments of this type.

There is a great deal of supplementary information that can be obtained from a desk study that should enable a preliminary identification of sources, pathways and targets and lead to the creation of a site risk profile.

A current Ordnance Survey map can provide much useful information, including local topography (in what direction would spilled material flow?), proximity of residents and local landmarks. They are helpful in establishing the routes of local watercourses (the routes of on site drains and ditches etc, which lead to watercourses will, of course, rely on site observation). Historical maps are a primary source of information on potentially contaminative previous uses. Although the process of obtaining such information can be undertaken directly through library sources, a number of organisations now offer a cost-effective and rapid map searching service.
The hydrogeological sensitivity of a site may well be an issue that requires detailed specialist study, however, a broad indication may be obtained by reference to Environment Agency publications or those produced by bodies such as the British Geological Survey (BGS). A useful starting point is the *Groundwater Vulnerability Map of England and Wales* (1992) which describes five sensitivity classes depending on underlying aquifers and their water resource potential along with soil classification (whose physical properties affect the downward movement of pollutants). This map is supported by more detailed (1:100,000) regional groundwater vulnerability maps produced by the Environment Agency.

Should radon be considered within the scope of hazards to be assessed, initial reference could be made to publications such as the *Radon Atlas of England* prepared by the National Radiological Protection Board (NRPB) in 1996. The Yorkshire and Humberside Regional Technology Network (1994) described a staged approach that involves moving from preliminary considerations to a desk study. The desk study was seen as an essential step to make any subsequent site investigation cost effective, in a process that could result in the curtailment of more exhaustive searches. Although contaminated land was the principal focus, the general approach in the report is applicable more widely.

The first step was to identify what is known about the site and liaise with external organisations including the following elements:

- Have any complaints been made to the local authority about the land or processes taking place on the land?
- What knowledge exists of the history of the site records or among staff?
- What local knowledge exists on the history of the land?
- Has the site been previously reclaimed, if so, by whom and for what purpose?
- Consider possible limitations of end use if contamination is present.

Moving on from the initial information, a desk study may then be undertaken, whose typical scope might include:

- establishing former use/uses of the land;
- establishing former use of adjacent sites;
- assessing likely potential of pollution from processes;
- producing list of possible contaminants;
- determining geological profile of land;
- examining local and regional hydrological regimes;
- assessing likely pollution of land;
- making a reconnaissance visit to site;
- assessing possible limitations on future use;
- identifying targets ie sites, activities, people, areas, habitats etc, which might be affected by anticipated contaminants;

- assessing potential for transport of contaminants between site and targets;
- re-appraising financial implications and risk of future actions; and
- identifying services beneath the site and their possible location (to be checked on site).

An excellent recent guide (Syms, 1999) has been produced to give guidance to surveyors, valuers and property developers when faced with land which may be contaminated. It describes a 'desk-study' type approach whereby typical industrial uses of land that may give rise to contamination are discussed along with advice on which actions might be required.

Property Transfer Risks

The route through which many organisations have been exposed to the process of environmental risk assessment in the first instance is through a property transfer assessment. This is an assessment of environmental risks and liabilities, usually made by the purchaser, in the course of an acquisition of sites and operations. Increasingly lenders, are also making an assessment of environmental liabilities a precondition for loans.

This type of project is sometimes called a pre-acquisition survey or audit. The current increasing interest in such work reflects US practice over the last few years and indeed much of the work conducted in the UK has been for US companies in relation to UK purchases. In the US, organisations can face very significant liabilities under legislation such as the Comprehensive Environmental Response Compensation and Liability Act (CERCLA) which allows the federal government to recover clean-up costs for contaminated sites from the owners and other parties. The average cost of cleaning up a site has been estimated at $20 million by the US Environmental Protection Agency. A later amendment of the legislation (SARA) allowed an 'innocent landowner' defence as a means of avoiding liability, provided 'all appropriate inquiry' had been made at the time of purchase. This had the dual effect of promoting further interest in pre-acquisition studies and a desire for standardisation (this would help to classify what 'appropriate inquiry' constitutes). Priznar (1991) quoted evidence that of around 5,000 environmental auditors in the US, approximately 75 per cent focused exclusively on property conveyance activities.

Property transfer studies are quite commonly known as pre-acquisition audits and while it is certainly true that they share many similarities with 'conventional' environmental audits, there are also some important differences. In property transfer work there is much less concern with current site practices and systems such as the presence or otherwise of a corporate policy. The focus is much more on the nature and extent of potential liabilities that might be inherited. Elements such as historic waste disposal practices are clearly of greater interest. A Phased 1-2-3 approach has become the norm for property transfer studies.

Phase 1 typically utilises a site walkover survey combined with historical and published information and interviews to assess the potential for releases of hazardous materials having occurred. Should Phase 1 suggest a release may have occurred, a Phase 2 study is initiated. Typically, a Phase 2 study would involve considerable sampling and laboratory analysis to determine the presence (or absence) of potential contamination indicated by the Phase 1 survey. This, in turn, could lead to a Phase 3 survey that would aim to define the degree and rate of migration of on-site contamination and its potential effects on human health and the environment.

Phase 1

Essentially a data gathering exercise which is documentary in nature, desk based and could include an analysis of site maps, records and other relevant material. The aim is to assess the potential contaminating nature of past uses on the site. This is usually complemented by a site walkover from an environmental specialist. One of the key aspects of a Phase 1 report will therefore be whether further work (ie a Phase 2 study) is required.

Research undertaken by the Construction Industry Research and Information Association (CIRIA) and supported by the DETR on the sale or transfer of land that may be contaminated indicates that following a preliminary investigation indicating the actual or likely presence of contaminants, the vendor may choose to:

- conduct further detailed investigation;
- remediate prior to sale; or
- sell as seen.

The types of issues that might then arise in negotiations arising out of more detailed investigation might include whether:

- the desk study work has provided sufficient information;
- there is sufficient time for more detailed site investigations;
- there is a mechanism for price adjustment/remediation should contamination assume significance after completion of the sale; and
- the parties can agree to joint instruction of a specialist for further investigation.

Conducting A Phase 1 Study

UK practice in relation to Phase 1 studies differs from the USA in that the review of site history and practice is more reliant on maps, directories and local authority information than property title searches. The site assessment, typically of one day duration is essentially similar, looking for indicators of potential problems such as soil staining, vegetation stress and the presence of USTs which could be inferred from vent pipes or fill points. Unlabelled drums or containers or poor housekeeping might indicate an increased risk factor. A typical scope for a Phase 1 Study would be as below:

Desk Study: Review of available information including:

- historical land use;
- planning permissions;
- ownership and lease arrangements;
- waste disposal records;
- discharge consents;
- geology, hydrology and soils;
- ecology and environmental sensitivity.

Site Visit: Site visit (typically one day) which will include visual inspection of the following:

- available documentation;
- site setting, surrounding land use and neighbours;

- visual impact, site tidiness and security;
- potential for air emissions, point source and fugitive;
- potential for noise and vibration nuisance and other nuisances;
- liquid effluent discharge arrangements, surface water, drainage sewers and groundwater;
- materials storage;
- underground storage tanks;
- hazardous materials such as asbestos, PCB, lead, radon and pesticides;
- evidence of land contamination;
- fire, spill and emergency protection;
- heating, power and water supplies; and
- evidence of uncontrolled waste disposal.

Report:

- executive summary;
- a brief description of the site;
- a statement of the environmental sensitivity of the location;
- identification and evaluation of potential risks and liabilities from previous uses typically based on a source-pathway-receptor approach;
- recommendations for further (Phase 2) investigation. If there are no recommendations for further investigation, this acts to confirm that, in the auditor's opinion, the relevant issues have been identified and no further action is considered to be necessary;
- identification and evaluation of potential risks and liabilities relating to the current operations in the context of their proposed future use;
- recommendations for any design improvements that may be advisable to ensure legislative compliance, reduce risks and generally ensure best practice; and
- appendices as appropriate.

NB: a Phase 1 audit does not typically include the collection of samples (eg soil) for subsequent analysis. Such requirements would be identified in Phase 1 and performed as part of Phase 2.

Property Transfer Studies - Outcomes

There is relatively little published data on the outcome of pre-acquisition surveys, although a US study (Oullette, 1990) indicated that 75 per cent of Phase 1 studies indicated no identifiable actual or potential problems. In 2 per cent of cases there was an end of interest and in 3 per cent the deal proceeded with or without resolution of an identified issue. In 20 per cent of cases a Phase 2 study followed, of which 12 per cent proceeded, 2 per cent terminated the deal and 6 per cent led to a Phase 3 study. Although the proportion of deals terminated on environmental grounds looks relatively small, it should be borne in mind that on the basis of a simple survey, a deal which may have been worth millions of dollars has fallen through. Not only that but the potential risks identified in the 20 per cent of Phase 1 surveys that lead to further investigation are likely to be used as a bargaining device by the purchaser.

A recent UK survey of environmental specialists undertaken by the author covering approximately 2000 surveys found that the most likely outcome following the audits was that the deal proceeded, either with no significant problems identified (27 per cent) or with some account taken of environmental issues eg an amended price or warranty (33 per cent) (Pritchard, 1998). In almost one-quarter of the cases the identified potential for significant liabilities was such that a Phase 2 study was initiated. In a small but nonetheless significant number of cases (4 per cent) the results were considered to have been a significant factor in the prospective purchaser backing away from the deal. It is interesting to note that 75 per cent of the audits in the US study revealed no apparent problem, a significantly higher figure than the UK. This could reflect the more routine nature of the undertaking in the US, whereas the studies have historically only been triggered in the UK where potential problems are already anticipated.

Phase 2 intrusive studies aim to confirm and quantify the extent of potential liabilities highlighted by the Phase 1 work. Of the total Phase 2 project outcomes 3 per cent led to the breakdown of the transaction, while 30 per cent indicated that the potential issues identified earlier were not significant. The most frequent result was that the deal progressed with some account taken of the Phase 2 findings – this is unsurprising given that the financial implications of liabilities should have become much clearer. In 19 per cent of cases detailed Phase 3 investigations were instigated.

The Phase 3 investigation would be conducted with the objective of defining the extent, degree and rate of migration of contamination identified in Phase 2. It should also contain sufficient information to evaluate potential impacts on human health and the environment as well as proposing remedial action alternatives. There is likely to be further soil sampling and analysis, although a large proportion of the effort is likely to focus on the nature of surface water and groundwater contamination.

Contaminated land remained the principal concern identified in the survey, however, respondents also said upgrade costs to ensure ongoing legal compliance of plant and process equipment had become more important. Other potentially significant concerns included:

- the presence of significant quantities of asbestos on site;
- noise problems, often difficult and expensive to abate;
- the presence of polychlorinated biphenyls (PCBs), most notably in electrical transformer oil;
- location in an area which has elevated levels of naturally occurring radon; and
- underground storage tanks – where leaks of oil, fuel etc, may potentially go unnoticed for long period.

A more detailed description of the survey methodology and results is given in Appendix A.

Cahill (1989) has considered the types of environmental factors that might constitute 'deal killers' in the US, including:

- site built over a landfill operation;
- the presence of large amounts of asbestos containing materials;
- high risk neighbours with evidence of leaks or spills;

- soil and/or groundwater contamination in combination with geological or hydrogeological characteristics that exacerbate problems;
- extensive PCB contamination;
- high radon levels;
- conditions representing breaches of legislation; and
- properties that are the subject of environmental litigation.

There is an obvious correlation, despite legal differences, in the types of issues that are of concern in the UK and US. Costs associated with the surveys can obviously vary considerably. However, the following reported figures (in 1990 US dollars) offered a reasonable guide (Oullette, 1990):

Phase 1 Study	$ 2000–13000
Phase 2 Study	$30,000
Phase 3 Study	$100,000 + (if groundwater cleanup is involved, costs might increase by a factor of ten)

Whereas it is clear that there are significant advantages for prospective purchasers in commissioning pre-acquisition surveys, the situation for vendors is not so clear. In a regime underpinned by 'let the buyer beware' the vendor may question the wisdom of undertaking any specific environmental work. Syms (1997) considers in some detail the case for and against site investigation and the disclosure of environmental information by the vendor, as summarised below.

Advantages:

- A baseline as to the condition of the site at the time of sale is established, safeguarding the vendor in the event of any future legal action. It may open up the 'selling with information' defence contained in the Environmental Act 1995.

- It enables the vendor to clean-up or contain any contamination in respect of which the vendor could remain liable.

- Where the site is to be redeveloped for a more sensitive end use, it enables provisions to be inserted into the contract for sale.

- It avoids the situation where the purchaser undertakes a survey, and then claims that the cost of treatment is far in excess of those which were previously anticipated, thus placing the purchaser in a far stronger negotiating position.

- Disclosure may dispel, or at least mitigate, any suspicions held by prospective purchasers, and reduce the transaction timescale. The findings of the survey may in any case be favourable.

Disadvantages:

- If the audit discloses adverse findings the vendor may wish to keep them confidential, but this may be a problem if prospective purchasers become aware that an audit has been undertaken and specifically request information.

- Purchasers with little or no experience in land transactions, or in dealing with contaminated sites may be 'scared-off' no matter how trivial the extent of actual contamination.

- There must be some concern as to who that information may reach, which may be addressed by the use of confidentiality agreements.

On balance, Syms conclude it is generally favourable for vendors to disclose information. It is also important to recognise the risk perception aspects associated with the identification of contamination. Petts (1994) studied the blight effects of the identification of contaminated sites to understand responses to sites that may be contaminated but which do not appear to pose immediate significant risk. Sources of (or influences on) blight and alarm were identified as:

- sudden identification of a previously unknown possible problem;

- lack of an identifiable responsible party;

- apparent lack of technical understanding by the perceived responsible regulator (in these cases the local authority); and

- the time lag between identification of a problem and action being taken.

None of these factors relate to the actual magnitude or the nature of the risk, emphasising the need to consider broader issues (particularly communication) to achieve effective risk management.

Property Transfer Standards

The innocent landowner defence in the US enabled a prospective purchaser the potential avoidance of inherited liability provided they conduct appropriate enquiry into the nature of the site, including potential contamination. There has accordingly been considerable effort expended by agencies in the US, including several states and the American Society for Testing and Materials (ASTM), towards generating standards for property transfer assessments, which would enable prospective purchasers to ensure they could satisfy the 'appropriate inquiry' defence.

The ASTM standards E1527-97 (1997) and E1528-96 (1996) provide guidance in the US on good site assessment practices satisfying the due diligence responsibilities in commercial property transactions. The two standards, developed by committee E50, are complementary. E1528 describes a transaction screen and E1527 a Phase 1 environmental assessment. The transaction screen involves the use of a questionnaire and can be undertaken by persons who are not environmental professionals. At the conclusion of a transaction screen, either no further enquiry may be required or it may lead on to a Phase 1 study. The Phase 1 environmental site assessment would be expected to be undertaken by environmental professionals.

The transaction screen questionnaire includes 20 questions that are put to the owner and to the occupants along with answers resulting from a site visit. The questions would apply directly to any site outside the US; subsequent sections relating to government records and historical sources of information would however need some adaptation for worldwide application. Appendix 2 in the standard shows helpful photographs of the type of feature that might be observed during a site visit. The approach taken by the screen is therefore capable of being widely applied, and as such is likely to be of interest to a broader audience including persons involved in property transfer outside the US, who want an indication of the nature of environmental liabilities.

The Phase 1 Environmental Site Assessment Process requires greater time and environmental expertise, involving a records review, site reconnaissance and interviews with local government officials. It also describes the following principles underpinning the assessment process, which emphasise the aspects of uncertainty.

- Uncertainty not eliminated – recognising that assessment protocols reduce, but do not eliminate uncertainty.

- Not exhaustive – the process attempts to balance the competing aspects of controlling times and costs and the reduction in uncertainty from further investigation. The overall aim is not to undertake exhaustive enquiry in all circumstances.

- Variable level of enquiry – not every property requires the same level of enquiry. The type of information required will be dependent on several factors, including the nature of the property and the risk tolerance of the person who commissions the work.

As with E1528, many of the elements of the Phase 1 study could be used directly in applications outside the US, and those of a US specific nature (eg legislation) are capable of being adapted to reflect different national regimes. This contrasts with, for example, other environmental audit packages where legal compliance elements are paramount. In property transfer the emphasis is on liabilities that are of a broadly similar nature in most countries, whereas compliance audits are likely to reflect closely the type and standard of environmental regulation of their originating country.

Contaminated Land

Contaminated land is one of the principal environmental risks associated with property. Viewed from an international perspective, however, different countries assign differing significance to the issue. In some such as the Netherlands and the US it is viewed as an extremely important environmental issue whereas in others, even in neighbouring EU states, it receives little attention. As with many issues there have often been a small number of highly publicised events which led to a rise up the political agenda, for example, Love Canal in the US and Lekkerkerk in the Netherlands.

In the UK incidents such as those highlighted in Table 3.3 taken from *Contaminated Land and its Remediation* (RSC, 1997) have been important. Notwithstanding these incidents the principal concern with contamination in the UK has arisen during the development process (reflected in the original regulatory framework). This is unsurprising given the long history of industrial activity in the UK coupled with considerable pressure for the reuse of land.

Ferguson and Denner (1994) indicated that risk management (in the context of contaminated land) is required to balance the need to safeguard human health *and* support the activity that will fund the remediation process. In the UK this is often through the redevelopment process, accordingly if clean up standards are set very high, this may act to discourage the whole development that funds the clean-up and favour the use of a greenfield site. These conflicts are exacerbated in recognising that the cost of risk reduction can be highly non-linear and that the level of public concern about

Table 3.3 *UK Incidents and Contaminated Land Sites*

Site	*Date of Incident*	*Description of problem*	*Costs / Implications*
Clarke Avenue, Loscoe, Derbyshire	March 1986	Explosion due to landfill gas migration demolished bungalow, badly injuring the occupants	£325–375k; continued difficulties with mortgage lending and insurance
Ilford, Essex	Late 1980s	Low level radioactivity found on former chemical site being redeveloped for housing	£10m; extensive remediation and delays to site development
Lumsden Road, Portsmouth, Hampshire	1991	120 families rehoused after Ministry of Defence development of 1950s/1960s constructed on former Royal Navy waste site found contaminated with asbestos and heavy metals	£6.12m; including rehousing, medical screening, demolition, remediation and rebuilding

contamination can be much higher than might be expected from an 'objective' analysis of the risk. The fact that the risk is involuntary and incompletely characterised or understood serves to explain the perceived risk that needs to be balanced against the cost of protective actions.

The UK approach to remediation has been to make the site suitable for use, ie clean up is only required to the standard necessary for the relevant end use. Accordingly a less sensitive end use, for example, industrial rather than domestic might be specified for a development. This approach contrasts with that of multi-functionality which requires clean up to a standard suitable for any end use (which is of course potentially more expensive). The Netherlands has traditionally championed this idea. However, this multi-functionality approach is under review (ENDS, 1997a). The Dutch Environment Minister was quoted as saying '*As the years go by willingness is waning to contribute more to soil contamination of the past*'. This is unsurprising since the government has been contributing 70 per cent of the £500million annual clean up expenditure and that the estimated number of contaminated sites has risen to 100,000.

The delayed changes to the UK regulatory regime involve implementation of section 57 of the Environment Act 1995 to add Part IIA to the Environmental Protection Act. This will require local authorities to inspect their areas and identify land that is contaminated. Contaminated in this context does not mean merely the presence of contaminants but the requirement that it is associated with 'significant harm'. This will require an assessment of risk utilising the source-pathway-receptor approach. Following designation of land as contaminated the enforcing authority will serve a remediation notice on the appropriate person. Utilising the polluter pays principle, the appropriate person in the first instance will normally be whoever knowingly caused or permitted the land to become contaminated. However, if this person cannot be found, the appropriate person becomes the current owner or occupier of the land.

The approach adopted in the UK has had the benefit of drawing on several years experience in the US (and indicates trying to avoid some of the problems). For example, the unsatisfactory (except to lawyers perhaps) situation whereby a large proportion of available resources is devoted to allocating responsibility to potentially responsible parties. The new regime underpinned by the Environment Act can be viewed as risk management in at least two ways, firstly on a strategic level it aims to complement existing remediation efforts linked to development by ensuring that sites falling outside the existing controls, but posing a significant risk, can be remediated. Furthermore, the tools developed for the regime range in complexity up to detailed site specific human health risk assessments.

It is likely that a risk ranking type exercise will be undertaken by local authorities (and increasingly by purchasers of property) based on a simple source-pathway-receptor concept to establish an initial priority list. The next stage up would be to compare measured contamination levels with predetermined guideline values. Where the contaminant levels exceed the guideline values this is likely to trigger further investigation. This could then lead to a site specific risk assessment with associated data modelling and analysis requirements. The (expensive) site specific assessment could also be used in the derivation of site remediation requirements.

The guideline values and the site specific risk assessment utilise a similar framework with default parameters (ensuring an overall acceptable level of environmental protection) being used to derive general guideline values and site specific measured or modelled parameters being used in the latter case. The Contaminated Land Exposure Assessment (CLEA) model developed by Professor Ferguson at Nottingham Trent University for the Environment Agency underpins this process. It draws on the approach pioneered by RAGS in the US, however, it has incorporated recent developments. For example, a particular concern in RAGS has been the use of 'reasonable maximum' point values for variables to input into the calculation of risk. Since a large number of variables is used the end result is a value which can be massively conservative in its estimate of risk. The increase in available computing power has now facilitated the use of Monte Carlo-based methods which run thousands of simulations using different values consistent with the distribution functions of the individual variables. This builds up a distribution function for the final risk estimate from which a 90 or 95 percent (or whatever is selected to be protective of health) upper limit value is selected. CLEA is also interesting in that (unlike RAGS for example) it takes account of natural attenuation processes by inclusion of a simple decay equation.

The new UK Guideline Values derived from the Contaminated Land Exposure Assessment model (CLEA) make a generic assessment of direct human health risks via different exposure pathways. Risks to groundwater and surface water are not included because of the highly site specific nature of these risks. The site use is assumed to fall into one of the following categories: residential with or without gardens; allotments; parks; playing fields and open space and commercial/industrial. The average daily intake may be calculated with an averaging period of 70 years (a conventional lifetime) or shorter periods (Ferguson and Denner, 1998).

Should a remediation notice be served on an organisation actions taken will be dictated by the legal framework. Nevertheless, it more likely that the issue of contaminated sites will need to be addressed in the context of development or sale/purchase. The most likely scenario for a manager with responsibility for contaminated land is that technical environmental experts will be appointed to determine the nature and extent of potential liabilities.

This initially involves ensuring that qualified specialists are appointed under appropriate terms. Site investigations can present problems in terms of access in the first instance. They might then create problems through disruption of underground services or even opening new pathways for contamination to exacerbate existing hazard. Finally it should be remembered that the investigation will always be a sampling exercise and cannot guarantee that all contamination will be detected. The types of contractual issues to be considered are shown in Table 3.4

Table 3.4 *Contractual Issues Requiring Allocation of Responsibility and Liability on Contaminated Land Projects*

Determining underground service locations
Obtaining necessary authorisations and licences
The health and safety of all on-site workers
Site security and protection of the public, including trespassers
Access to all operational areas, eg sampling and monitoring locations, excavation and processing areas
Waste disposal, effluent discharges and environmental protection
Unforeseen conditions (eg more contaminated material than expected)
Re-instatement of trial pits, boreholes, fencing, landscaping and hard cover
Vandalism or damage of contractor's plant and installed monitoring equipment
Accidental damage to property or injury to third parties
Abortive boreholes and non-functioning monitoring and treatment installations.
Restrictions to site operations – impedance of access and enforced standing time due to client's own operations, limits on equipment that can be used (eg intrinsically safe electrical gear and use of petrol engines) and permitted working times
Delays – slow provision of approvals, information and payment, project over-runs due to bad weather, plant breakdowns and late delivery
Causing spread of contamination during disturbance of contamination soils, water or sludges, including installation of boreholes.

Source: A Quality Approach for Contaminated Land Consulting, CLR Report No 12 Department of the Environment, 1997.

Following appointment and investigation a report will be produced and presented to the manager. It will almost certainly contain details of soil and groundwater sampling protocols, measured concentrations of individual contaminants in soil and water (hopefully with confidence limits). There may have been computer modelling to predict the extent of off-site movement of contaminants in groundwater.

Further actions could then include the following:

- *No (or little) Action Required:* the levels determined were sufficiently low to conclude that little risk is posed by their presence. A word of caution: what may be considered acceptable at the time of the report might not always remain so.

- *Remedial Action Required:* the levels observed were sufficiently high (eg exceeded Guideline Values) to conclude that remedial action is definitely required. The next decision would then need to be: what type of remediation is appropriate and how far should it go? Historically in the UK, two types of remediation were employed in the great majority of instances; transport off-site of the contaminated material or covering the affected area with imported clean material, concrete or the like. It is important to recognise that although these approaches might well be seen as appropriate in many circumstances, they are not no-risk options. When the risk source remains on-site it clearly has the potential to migrate and cause further problems. Removal and disposal off-site involves risks of its own; these will be the transport of hazardous material and then, almost certainly, deposition in a landfill site. The risk source is not destroyed, but is transferred to an area where it can be better controlled.

In future there is likely to be a greater insistence on treatment of the contamination which could include: in-situ bioremediation; incineration; immobilisation; electroreclamation; and inserting containment walls or screens. Complementing this is an improving understanding of natural degradation processes which enables better prediction of decreases in contamination. Interest in this process of 'natural attenuation' has been the subject of considerable recent research in the US, not least because it provides a better, defensible case for no or low level intervention. It also provides comfort in giving an indication of future reducing risk levels as natural degradation progresses. The cost associated with various remediation options is likely to vary significantly as shown by the figures in Table 3.5. There are specific technologies that are more appropriate to certain types of pollutant, although there may be difficulties in deciding between the competing claims from various proprietary approaches.

Regulatory bodies such as local authorities or the Environment Agency will be involved in the assessment and remediation process. From the point of view of advisors, it is very unlikely that they would provide simple clean-up standards immediately. It is much more probable that the burden of proof will lie with the organisation to prove to all interested parties that it has behaved responsibly and complied with legal requirements in proposing its own scheme. The combination of several options, each with a potentially significant uncertainty, makes risk-based approaches useful in the selection of remediation options. In some instances the hazard may constitute such a small risk that it may be left in place, whereas all remediation options are likely to involve significant capital expenditure.

In some cases it may be possible to play safe and remediate back to a trivial risk level. However, when faced with a large site (or several sites), each of which might require heavy capital investment, this might not be appropriate. The manager should aim to have a sound defensible basis on which to base such decisions. An appropriate approach for the executive could be decision analysis, which specifically aims to take complex technical data with its inherent uncertainties and translate it into a form which can be used easily by the manager. The complex overall decision is broken down into smaller decisions which are then recombined.

Table 3.5 *Summary of Remediation Techniques*

Technique	*Method*	*Examples*	*Cost*
			(1993 order of magnitude)
Containment	surface capping	tarmacadam, synthetic liner, imported soil	£35–75/m^2
	vertical barriers	diaphragm walls impermeable walls sheet pile walls	£60–£180/m £25/m £20/m
	horizontal barriers	jet grouting	£280/m^2
Removal	off-site disposal	landfill	£5–30/tonne
Treatment	Replacement of soils treated	soil washing incineration thermal stripping	£80–120/tonne £400–1,200/tonne £40–350/tonne
	Reduce contaminant sources	pumping & treatment of groundwater biological treatment	Varies with method £16–80/tonne
		chemical treatment vapour extraction	£135–350/tonne £25–75/tonne
		vitrification cement treatment 'lasagne' multiple-layering treatment	£250–275/tonne £50–100/tonne (not yet viable in the UK)

Source: *Tackling Contamination:Guidelines for Business to Deal with Contaminated Land,* Confederation of British Industry, 1994.

In the first phase, sources of impact are identified and include the potential targets on which the hazards might act and the pathways to the targets. For example, off-site migration of a contaminant in drinking water might affect a nearby well used for drinking water. Different scenarios are accounted for by branching out the sources to account for different possible outcomes. Much of the utility of this approach comes from asking the right sort of question of the technical specialist and requiring them to include uncertainty estimates. In many cases the information necessary to make a decision between remediation options will be less than is required to characterise fully all aspects of the nature of contamination. This approach can therefore help by ensuring that the costs and efforts associated with characterising the problem are restricted to what is needed to make management decisions. The commissioning of unnecessary sampling, analysis and modelling (even though they may be interesting exercises for the technical specialist) are thereby avoided.

The raw data on outcomes, each with their associated probabilities, can then be recombined to give overall cost values with confidence intervals for each risk management strategy considered. It is also likely that the manager will be more concerned with non-financial consequences such as public relations and a discussion of relevant factors might be included with each management strategy under consideration. In summary, what is required is a decision-focused approach that uses available expert information and takes account of uncertainties to provide non-specialist managers with identifiable options based on available technical data.

Developments in information technology will facilitate the manipulation and management of data for the risk assessment of contaminated land projects. This will need to accommodate even more than the site specific three–dimensional distribution of contaminants and potentially complex geological variations. The situation is also time dependent since contaminants move under gravity and other gradients. Chemical and biological processes may result in breakdown or changes in chemical state and the climate may also be influential (eg evaporation may increase in warm weather or greater evolution of ground gas may occur during periods of low atmospheric pressure (Nathanail, 1995).

Geostatistical techniques offer potential for dealing with the results of contaminated land site investigations (Upton, 1998). Investigations have often historically been driven by the highest determined point values (which in themselves have received extra attention because of odour, discoloration etc). Using these values is likely to introduce significant conservatism into risk assessment protocols and a value for the average bulk source concentration might be more appropriate. Modern geostatistical techniques can therefore be usefully applied to estimate means and standard deviations of datasets. This will also permit identification of whether there is a single (hot spot) or multiple sources of contaminants and identifying where additional samples should be taken to achieve adequate dataset confidence. Block source values are of course likely to be significantly lower than the highest point source values.

Fire Related Environmental Risks

Fire has rightly been recognised as an important business risk for centuries. In addition to the obvious safety aspects, it can impact in areas such as business interruption, as well as causing major financial losses. The importance of potential environmental impacts has only been recognised more recently. Obviously there are potential impacts associated with the (highly visible) gaseous and particulate combustion products. However, the consequences of contaminated firewater affecting watercourses are perhaps even more serious. The importance of environmental concerns was emphasised recently when a waste company had its site licence suspended for failing to co-operate with the Fire Brigade in controlling site fires (Environment Business, 1999b).

Measures taken to reduce the likelihood of a fire starting will obviously be most desirable, however, addressing environmental aspects will require consideration of appropriate contingency arrangements. Consideration of the pollution risks to water from contaminated firewater should therefore consider both the sources of release of dangerous substances in the event of a fire and the potential consequences that arise from:

- the properties of the substances present on site, for example, toxicity, flammability etc, and the potential for adverse reactions between them;

- the environmental behaviour of the substances such as solubility, volatility, persistence, potential to bioaccumulate in the food chain and environmental toxicity. Products formed as a result of both complete and incomplete combustion should be considered. These may be more persistent and harmful than the parent compounds;

- information about the location, inventories and process conditions, with reference to facilities such as bunding, entry points to storm and foul water drains, protective equipment associated with them, and the routes of water flows to containment and treatment plants; and

- information about the type and vulnerability of relevant water resources which will include:

  – sites within river catchments above public water supply intakes and/or reservoirs (time of travel to intake, the importance of the intake and other similar factors may need to be considered);

  – sites in close proximity to aquifers or close to watercourses that flow into and can contaminate aquifers, particularly those used for public water supplies; and

  – sites which pose a threat to waters of particular ecological value such as sites of special scientific interest (SSSIs).

Means to minimise the risk from firewater normally comprise arrangements for containing the run-off. Guidance is available in the HSE Guidance Note EH70 *The Control of Firewater Run-off from CIMAH Sites to Prevent Environmental Damage*. (It should be noted that the CIMAH Regulations have been superseded by the similar COMAH Regulations, which explicitly require consideration of off-site environmental impacts).

For example, it is a sensible precaution to store environmentally hazardous chemicals in bunded areas. Ideally, the capacity of the bund would be sufficient to take the entire inventory of the chemicals stored within it, plus the foam or water from fire protection activities, but this may not always be achievable. A bund sized to take 110 per cent of the entire chemical inventory might not provide sufficient capacity to prevent firewater run-off. However, it may provide sufficient time for emergency, temporary containment measures to be brought into operation.

Permanent provision for the containment of large quantities of firewater run-off, typically several thousand cubic metres and above, may be achieved by lagoons and tanks. These can be remote from, and may serve, several storage areas, receiving the run-off via a gravity flow or pumped drainage system. Alternatively, sacrificial areas such as natural depressions, or car parks might be used. An important aspect to consider is the permeability of the ground and how to minimise soaking of the pollutant into it.

Recent pollution prevention guidelines (Environment Agency, 1999) outline the factors in deciding capacity for remote containment:

- primary capacity to include the capacity of all tanks in multi-tank installations;
- rainfall volume allowance should include, for a ten year return, eight days rainfall prior to the incident, in addition to other aspects;
- allowance for extinguishing and cooling water;
- freeboard allowance for foam of not less than 100mm; and
- allowance for 250mm for dynamic effects such as the initial surge of liquid.

The COMAH Regulations and the Agency guidance are aimed principally at high hazard sites, however, the same principles can be applied to any site, and might be integrated into routine fire inspections. In a simplified risk assessment the first step would involve the identification of environmentally hazardous materials on a site (ensuring that products with high BOD potential and solids or dusts which could give rise to high levels of suspended solids are included). Secondly the consequences of a fire should be envisaged, ie where would the firewater run, obviously the destination of local drains and ground slope are the fundamental factors here. Finally, what contingency measures can be put in place, eg can people at the destination of the firewater (eg a water treatment works) be contacted easily, can water be retained on-site eg by closing a stop valve on the drainage system, diverting to a car park or switching off an automatic sump pump. In an emergency such as a fire there will be no time to assess these items that could easily be identified in only a few hours preparatory work.

It is also worth noting the potential adverse impacts associated with halon-based fire fighting systems. Their inadvertent release can lead to significant damage to stratospheric ozone. This is an issue to consider in terms of controls exerted as these systems are decommissioned leading up to 2003.

CHAPTER 4
FINANCIAL RISK TRANSFER

One of the principal tools for managing risks has traditionally been the use of insurance. The experience of dealing with environmental risks via insurance has, however, not been universally successful. The view from the insured has often been that an insurer would require an expensive survey before insurance could be obtained, if contamination was found then cover would be refused, and if the site was clean – why was insurance needed anyway? From the insurer's point of view issues such as contaminated land posed particular difficulties, with an unclear regulatory framework behind a risk where realisation potential and cost were both very difficult to estimate. In addition, the damage might not be evident for a long period and it might be difficult to identify when it happened in the first instance. The unhappy experiences of US Environmental Impairment Liability (EIL) underwriters in the 1980s served to underline this caution.

There are of course elements of truth in these views. This chapter will therefore consider some of the potential pitfalls associated with environmental insurance and how they might be overcome. Recent developments seem to indicate that positive progress is being made, and there would seem to be a demand for new approaches. In the US the environmental insurance market place was considered to be generating about $1.1 billion in premium income in 1997 (Hovey, 1997). Interestingly in this situation 20 per cent annual reduction in prices were thought to be part of a continuing trend. A recent survey on risk financing and insurance conducted by Aon (1999) placed environmental risks as top concern in relation to the availability of insurance cover, which would seem to indicate that there is demand in the marketplace.

Loose (1998) noted recently that insurers are taking a more flexible approach in adapting policies to meet site or company requirements. Specifically, that insurance products may now cover a wider range of environmental liabilities and for longer periods. Cover which can be transferred on sale of a site was also seen as advantageous. Set against this, some ongoing concerns were highlighted, including the claims made basis of cover and the requirement that, in the case of mandatory clean up, the insurer may retain the right to have the remediation work undertaken by its own consultants.

The view that traditional insurance solutions are no longer seen as offering adequate protection as the nature and scale of risks facing organisations is subject to rapid change is reflected in the opening sentence from a recent *Financial Times* survey (FT 1998) *'Rogue trading, failed product launches and damaged reputation can inflict greater harm on an organisation than the strongest earthquake or hurricane'*. Discussions of newer initiatives in this chapter will be complemented by a discussion on the extent of cover available on existing insurance policies. There is also reference to the utility of insurance in relation to specialist advisers (principally Professional Indemnity (PI) insurance).

Finally, it might reasonably be asked whether insurance or other financial mechanisms contribute to overall environmental improvement. Essentially they are vehicles for sharing and transferring risk, which is a supporting role, but can be nonetheless vital in facilitating progress. For example, the possibility of large and unbudgeted remediation costs associated with brownfield sites could discourage development and lead to increased pressure on greenfield locations. A situation which could be counteracted by having appropriate insurance cover in place. Risk transfer can therefore faciltate such actions by removing the potential for project-threatening adverse outcomes.

It is instructive to start with some fundamental insurance principles that will underpin subsequent discussion.

Firstly, any insurable loss must be 'fortuitous'. Taking a motor vehicle example one would not experience an accident and then expect to be able to approach a broker after the event and gain cover relating to the accident. Insurance is used where there is a risk of an adverse event, not where it has already happened (or indeed where there is no possibility of the adverse event arising). It therefore follows that if the nature and associated cost of contaminated land was already known it is not likely to be insurable. However, as we will see, other insurance and risk finance products might be usefully applied, for example, in the remediation process.

Secondly, any ambiguity in insurance policies is likely to be judged in favour of the insured, hence if some item is not specifically excluded from a policy it will usually be deemed to be included. This is particularly relevant to the historical amendment of public liability policies described below.

Thirdly, there are a number of concepts that for historical or other reasons underpin the writing of policies. There is a distinction between first (ie the insured's property) and third party damage; separate policies are usually in force for these aspects. There is also a distinction between 'sudden' and 'gradual' pollution. Legal interpretations of the distinction, particularly in the US, have varied somewhat, leading to potential uncertainty. Indeed, it is often surprising to those outside the insurance profession to realise that the extent of cover has often been established only by resort to the courts.

The following example taken from *Environmental Liability Insurance* (Lockett, 1996) demonstrates some of the difficulties with the 'gradual' definition, particularly the wide interpretation that could be put on the wording. An explosion attributable to gas emanating from a landfill site would initially seem an obvious 'sudden' incident. However, in considering this case the root or underlying cause would be sought, which would then be attributed to the 'gradual' deposit of waste in the landfill, or the gradual decomposition of the waste in the landfill (or gradual seepage of methane in the space prior to ignition). The recommendation could therefore go to the underwriters to exclude the claim, opening the door for the insured to challenge the policy wording in court.

Pollution in general presents particular problems in trying to fix the time it occurred and also in proving the connection between the cause (ie the pollution) and the effect (eg impaired health of claimants). Furthermore, in areas where a range of industries has been operating over a substantial period of time, identifying who has been responsible may not be simple; for example, pollutants can diffuse through groundwater making it difficult to identify a single source.

Finally, in order for the insurance to remain economically viable there has to be some means of assessing the likelihood of a risk being realised and its financial consequences, in order to set a premium (particularly where the financial claims might be very high). The withdrawal of insurers from some types of insurance reflects this basic problem. Operators of certain hazardous processes in Germany have been obliged by law to obtain environmental insurance, however, the reluctance of insurers to offer the relevant cover has caused problems with this requirement. Another approach in some countries such as France and Italy has been to set up 'pools' whereby groups of insurers come together to pool (and hence share) their risks. It should be noted that we are principally concerned with the harmful effects that industrial and commercial operations may have on the environment. Therefore, the impact of environmental agents (hurricanes, storms etc) on the organisation are not considered, (although they are likely to be insurable).

The starting point for any assessment of environmental insurance should be to consider what cover may be in force already through existing policies (ie those which may not be specifically environmental, but have not completely excluded environmental risks).

In the US in the 1970s insurers started to exclude gradual pollution from their public (general) liability policies. An increase in demand for environmental impairment liability (EIL) cover was then met by an increase in the number of policies on offer. Many of these programmes failed and there was subsequently a collapse in the number of insurers in the market. A number of explanations have been proposed for the problems experienced by insurers with EIL, including the following (Metelski and Gulledge, 1989):

- broadly accepting accounts by industry groups rather than on a facility-by-facility basis;
- insuring high risk industries;
- providing unlimited retroactive coverage and defence costs;
- not using personnel capable of adequately assessing environmental risk; and
- providing blanket coverage for owned sites and liabilities arising from non-owned waste disposal sites.

In the light of these difficulties the caution exercised by insurers is unsurprising. In the UK prior to 1991 there were no specific pollution exclusions on public liability policies, however, concerns lead to adoption on wordings such as those shown below, which limited coverage to sudden, unintended and unexpected events.

ABI suggested wording for sudden, unintended and unexpected pollution cover:

A. This policy excludes all liability in respect of Pollution or Contamination other than caused by a sudden identifiable unintended and unexpected incident which takes place in its entirety at a specific time and place during the Period of Insurance.

   All Pollution or Contamination which arises out of one incident shall be deemed to have occurred at the time such incident takes place,

B. (Aggregate Limit)

C. For the purpose of this Endorsement 'Pollution or Contamination' shall be deemed to mean:

   1. all pollution or contamination of buildings or other structures or of water or land or the atmosphere; and

2. all losses or damage or injury directly or indirectly caused by such pollution or contamination.

Limitation on cover means a potential market has existed for products with gradual pollution cover, however, uptake has been relatively low for a number of reasons. Those insured may feel that they already hold adequate environmental cover even with a limited PL policy and factors such as the following are perceived to be concerns:

- high premiums;
- requiring the insured to pay a considerable initial sum towards any claim (equivalent to an excess on a domestic motor policy); and
- requiring significant upfront costs for a detailed site survey. Considerable technical skill is also required of the underwriter.

New insurance policies are likely to be written on a 'claims made' basis, which requires a claim to be reported during the policy period. This provides insurers with more control than the 'occurrence' policies under which a claim can be made at any time as long as the occurrence happened during the policy period. There might also be extensions of policies currently offering third party coverage so that own site clean up is addressed. Whatever cover is obtained it is likely to involve a negotiation process to tailor policies to individual needs.

Notwithstanding the concerns and issues highlighted above the availability of new insurance products does indicate positive progress. Insurance products offered recently in the UK include.

- Insurance capping of contaminated land remediation costs (Steward, 1996). This aims to remove concerns for developers of brownfield sites about unforeseen rises in the cost of remediation works.
- Certification for contaminated land cleanup (ENDS, 1997b). Following assessment of a site a certificate (transferable to third parties) is offered which guarantees that cover can be obtained from a particular insurer.
- Insurance of underground storage tanks
- Land Sites Policies (Croner, 1997), During property transactions concerns may arise regarding historic contamination. With appropriate information, environmental insurance might be available prior to purchase.

There has been considerable recent discussion on the 'convergence of capital markets' whereby the clear distinction between insurance and other financial markets is becoming less clear (Business Risk, 1999). It has been suggested, for example, that capital markets could provide solutions for property owners addressing environmental risks (Writer, 1998) in at least two ways:

1 pollution risk could be converted into a financial commodity allowing it to be traded like a stock or share;

2 purchasing extra financial security by linking insurance to debt or share capital.

The attraction of this approach for catastrophe insurance is easily understood when one considers that Hurricane Andrew in 1992 caused $18billion worth of damage. This is obviously a very significant amount for insurers but compared to the daily $133billionn fluctuation of the capital markets it is quite small. Access to much larger markets would obviously facilitate future risk transfers. Insurance type risks might be attractive to capital markets because insurance claim events such as earthquakes are independent of stock market events. Property catastrophe risk (earthquake, hurricane and hailstorm) has been funded since 1992 as options and futures through the Chicago Board of Trade. Bonds are sold that pay interest; under certain specified conditions, however (for example, in the case of substantial earthquake losses) the issuer may be allowed to default on the interest on the bond, in some issues the capital of the bond may be at risk too. Hedging or speculating can be made on the level of claims based against an industry index. Borden and Sarkar (1996) showed an example of how a 'call spread' might be comparable to buying a layer of insurance. In this strategy the purchaser buys the right to sell and buy assets at certain prices.

In terms of environmental risks these developments could mean that a greater variety of financial instruments are available, particularly where these are linked to very large potential losses. Moreover, it is also likely to facilitate hybrid risk finance products. In contaminated land remediation this might involve some mechanism to raise the initial funds combined with an insurance element to protect against unanticipated contamination or other cost overruns.

Bennink (1998) has recently discussed a Californian US Superfund site where the potentially responsible parties (PRPs) are setting up a fund to cover the cost of site remediation as well as providing finite risk insurance to cover the cost of overruns, delays etc. These overruns might arise from discovery of additional contaminants, greater migration of known contaminants or change in regulatory standards. Additional third party liabilities that might arise include litigation from surrounding communities and exacerbation of existing environmental conditions.

Merkl and Robinson (1997) undertook a risk analysis utilising a database containing details of 10,000 US cases. This was done to understand the nature of risks at a site that was a candidate for contaminated land remediation and to assess the potential for risk transfer through insurance and related products. The database contained details of clean ups and their associated costs (including legal costs) and resulted in the following classification of risks:

Uncertainty over the cost of remediation: reflecting the fact that clean up costs are rarely easy to predict.

Liability arising from the remediation process: the act of remediation may cause a new release (eg by creating a new pathway to an aquifer) and trigger a third party claim.

Unexpected consequences on third parties: damage may arise to third party health or property.

Undetected Contamination: further contamination is uncovered in the course of the remediation process.

Post remediation clean-up liability: some contaminants are likely to be left in-situ which could give rise to a third party claim or be subject to future, more stringent clean up requirements (a change of use might also trigger this aspect).

The results from their study are presented in Figure 4.1 (reproduced with permission) in the form of cumulative distribution functions. This form of presentation lends itself to the assessment of risk transfer options since we can easily visualise, for example, the likelihood of post remediation clean-up liability costs not exceeding $6.8million is 95 per cent.

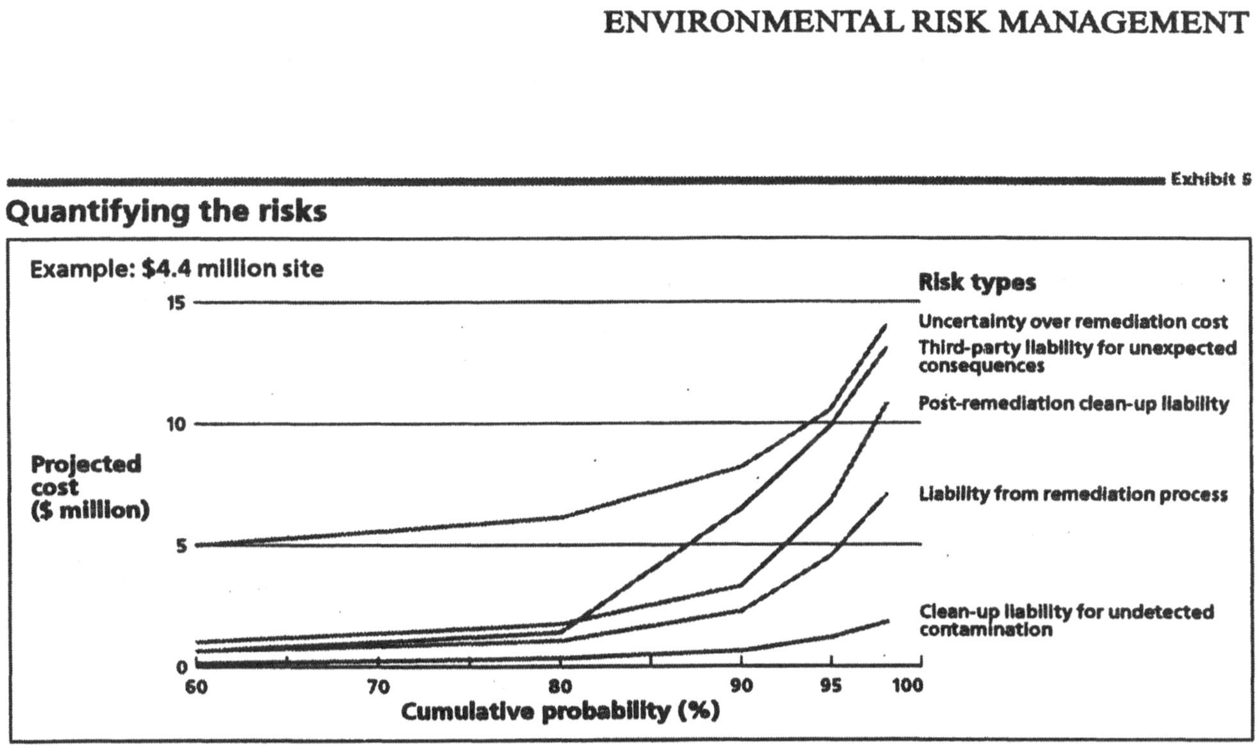

Figure 4.1 *Contaminated Land Remediation Costs*

The results demonstrate how the magnitude of these unplanned liabilities is relatively low (the projected cost levels in most cases do not start rising steeply until at least the 90 per cent cumulative probability level) but can be very significant at a small number of sites. Insurance cover might then be sought depending on the risk appetite of the purchaser. A risk averse organisation might seek cover for any liability greater than $1million, alternatively other organisations might feel comfortable with cover for the more unlikely event of significant cost overrun (say $10-15million).

A separate but related insurance issue is the insurance cover that may be carried by specialist advisers such as environmental consultants. For example, professional indemnity insurance provides cover against claims arising from errors, omissions or negligent acts. Many of the points raised in considering EIL cover are also relevant, such as the distinction between 'gradual' and 'sudden and unintended' cover. Accordingly, when asking a contractor about the insurance they carry the level of cover and the extent to which environmental risks are addressed is relevant. Of course the fact that an organisation carries PI for example, does not mean it is necessarily better than one which does not. The company may have taken a decision in the light of premiums and confidence in avoiding claims that they would carry the risk internally. Conversely, the fact that an insurer has assessed a company to be suitable for PI should give additional reassurance to the procurer of their services. A survey of environmental consultancies indicated that approximately 85 per cent maintained PI cover, however, less than one-half of these included gradual pollution damage (ENDS, 1995).

Insurance Risk Assessment

The assessment of environmental risks undertaken by insurers is basically similar to the process described earlier in the property chapter. The scope and cost of the assessment can be even more significant, in the sense that an unfavourable result may mean that the desired cover is not obtained. There is then likely to be the feeling that the expense involved is wasted.

The cost for an insurance environmental risk assessment was quoted in 1992 as a range from £1,500 for smaller sites, up to £5,000 – £6,000 for larger sites (Cook, 1992); this report also noted that only approximately 30 per cent of the sites surveyed up to that date were deemed insurable by insurers. It is unsurprising therefore that attempts are being made to control these upfront costs. Increasing use of desktop studies and information already available (eg through routine property surveys undertaken by insurers) should facilitate this process.

The identification and categorisation of risks also enables appropriate risk management measures to be taken to avoid or reduce risks, or identifies mitigating steps that can complement insurance cover (or indeed might be required as a prerequisite for cover). The increasing implementation of formal environmental management systems such as ISO 14001 should not only enable easier assessment of environmental risks but should also allow the demonstration of effective risk control.

A recent publication from the Association of British Insurers (ABI, 1998) aims not only to provide guidance to insurers, but also to increase awareness of the need for adequate risk management. A tiered assessment approach is proposed with a few simple questions recommended for proposal forms and a slightly more detailed questionnaire to be triggered as required.

An approach called the Zurich Hazard Analysis has been described for dealing with risk assessment and insurance issues associated with currently operating hazardous waste facilities (Schroeder, 1990). This methodology initially involves scope definition; identifying the systems or parts of systems to be analysed. The hazards are then identified and catalogued along with possible causes and potential effects. Each hazard is then assessed in terms of probability of occurrence and severity of effect. The production of a probability/severity grid for each hazard is called the risk profile and provides a management tool for risk reduction.

Four categories are used for the severity of occurrence of a possible effect:

1 *Catastrophic*: death, total disability, and widespread environmental damage.
2 *Critical*: severe injury with partial disability, local environmental damage.
3 *Marginal*: injury, temporary local environmental impairment.
4 *Negligible*: minor injury, little environmental disturbance.

The categorisation of probability of occurrence uses a scale with six levels as follows:

1 *Frequent*: often experienced or likely to occur frequently
2 *Moderate*: experienced or occurring several times
3 *Occasional*: sometimes experienced or occurring
4 *Remote*: may be experienced or may occur
5 *Unlikely*: unlikely to be experienced or to occur
6 *Impossible*: practically impossible (= lower limit)

The six hazard cause levels are scaled on a vertical axis and the hazard effects categories on the horizontal to give a risk profile grid as shown in Figure 4.2

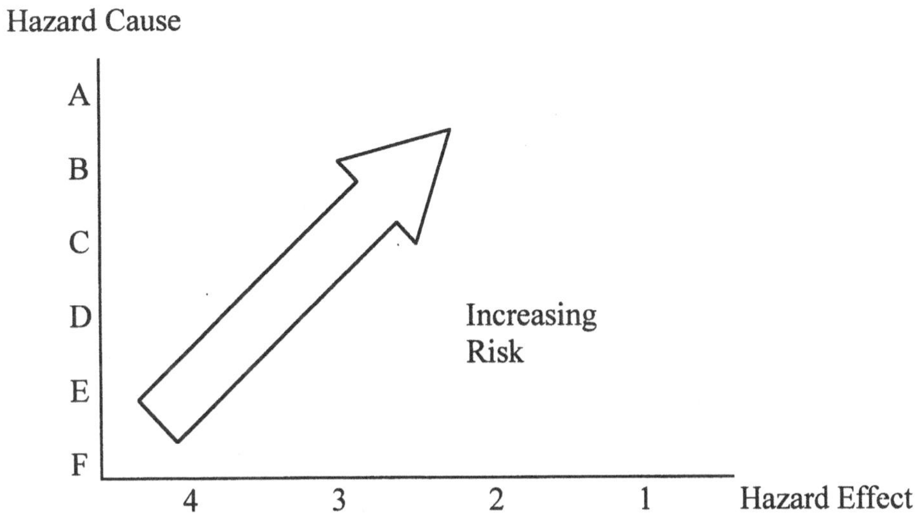

Figure 4.2 *Zurich Hazard Analysis*

CHAPTER 5
RISK MANAGEMENT INTEGRATION

The overall aim for organisations in managing risks is to ensure that they are built into the fabric of corporate decision making. In dealing with environmental risks an important issue then arises – where does environment fit into corporate processes? Several potential models are considered here with a particular emphasis on how environment might be integrated to avoid duplication of effort and ensure concerns are embedded within routine processes. Of course the models discussed cannot be regarded as prescriptive and the individual corporate context will determine which might be the most appropriate. A key factor is to ensure that environment is considered a strategic issue and receives appropriate senior level attention. This highlights some of the strengths and potential weaknesses of the formal environmental management standards such as ISO 14001 which are discussed here. They clearly offer an effective model for environmental risk control, yet care must be exercised to ensure that environmental concerns are not isolated and considered apart from other business concerns. For example, third party assessment can give credibility to a management system but when certification has been achieved the temptation to consider that the issue has been 'dealt with' and less worthy of board level attention must be resisted.

Recent concerns about the risks faced by organisations and their means of control have lead to guidance from a number of bodies such as the Institute of Chartered Accounts (ICA, 1999). Their recent publication provides guidance for UK listed companies on the implementation of appropriate internal controls to confirm that procedures are in place for evaluating and managing the key risks faced by companies. Accordingly a number of business risk assessment and reporting models have been produced. These concerns have arisen because of high profile losses such as those suffered by Barings, nonetheless there is often an acknowledgement that environment is one of the 'new' risks a company must manage (with Brent Spar often being cited).
The appendix to the ICA document presents questions that the Board may use in assessing internal controls. Environmental is identified as a potentially significant risk that should be identified and assessed on an ongoing basis. It was also suggested that an assessment be made on the effectiveness of communication to employees on what is expected of them in areas including environmental protection.

In an earlier publication (ICA, 1998) the ICA proposed that companies with publicly traded shares should lead moves towards enhanced reporting of risks in annual reports. In identifying and prioritising key risks they recognised that non-financial risks (ie those that do not relate directly to monetary assets but still might have financial implications) could include liabilities due to third parties arising out of environmental damage caused by a company's operations.

There is also likely to be increased interest in quantification of the financial aspects of environmental risk arising out of FRS12 (ASB, 1998). This standard sets out requirements for companies on the recognition and disclosure of provisions and contingencies (including environmental risk).

The business risk models that have been produced are usually directed towards functions such as internal audit aiming to provide boards with the means to assure shareholders that business risks are being effectively assessed and controlled. Environmental issues should form part of this strategic risk process, however, it is very unlikely that these processes will be run by environmental specialists. The challenge being faced is therefore to ensure that environmental concerns are considered within the scope of the review and that suitable mechanisms are in place to screen concerns and call in specialist expertise as required. A simple review checklist is presented to support a strategic issue screening process. Integrative initiatives are also considered in the context of links with health and safety. Considering these issues together is sensible given the same materials are likely to have both health and safety and environmental concerns.

The simple risk management model shown below in Figure 5.1 is not only consistent with the environmental paradigms described in this publication, but also underpins the broader business risk initiatives.

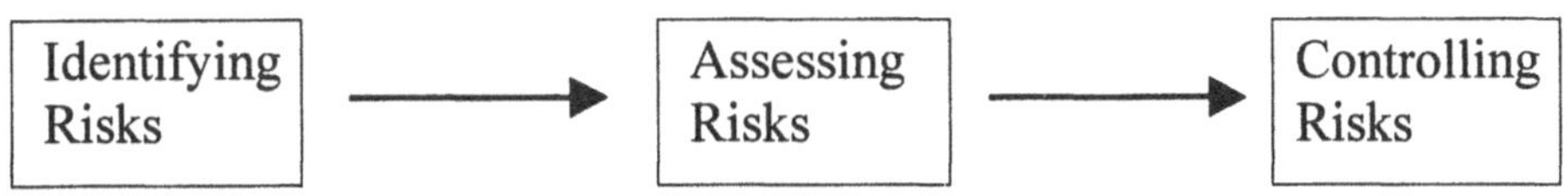

Figure 5.1 *Risk Management Process*

This model starts, from an initial mechanism for identifying risks through to some means of assessing which of these are important. Finally, management actions need to be instituted to control those identified as important. The risk management approach is also fundamentally consistent with the strategy described in environmental management system standards such as ISO 14001, which are discussed later.

It is also consistent with the approach described in the *Guidelines for Managing Risk* in the Australian Public Service (APS, 1996).

Key messages presented in these guidelines included:

- risk management is everyone's business;
- risk management is part of business as usual; and
- the process for managing risk is logical and systematic – and should become a habit.

This report addressed risks in general and identified environment (and community) as possible areas of impact.

Risk Models and Strategic Profiling

An *Economist* report (1996) has indicated that a more comprehensive model of business risk assessment and control is required. It cited reasons for this such as problems with financial derivatives trading, but also quoted 'confusing and conflicting environmental laws and regulations'.

The key elements in their proposed approach involved:

1 An integrated approach – to reflect the pervasive and interrelated nature of business risks.

2 Use of a common language for business risk (particularly pertinent in the case of environmental risks).

3 A control structure to anticipate and control risks – (which parallels the corresponding elements in ISO 14001).

4 Adopting a process view, emphasising horizontal linkages that cut across business functions.

Types of risk identified included:

- catastrophic economic loss;
- business interruption;
- loss of reputation;
- regulatory risks; and
- reducing cycle time also increases risk.

Environmental risks could conceivably impact in all these areas, although the historical lack of catastrophic economic losses has meant a low profile for this area.

The language used through the risk management process must be consistent and capable of being understood across the business. Three broad headings of risk were described as a useful means of risk classification: – environment risk, process risk and information for decision making risk.

Environment: (NB not environmental) relates to external factors outside the direct control of the organisation. This could include the actions of competitors and regulators as well as market forces.

Process: relates to ineffective business processes. This would mean that the process is not operating as intended or is not aligned with overall business aims. It might also mean that the operation of the process exposes the organisation to risk.

Information for decision making risk: this arises when information used for making business decisions is inadequate, incomplete, and irrelevant (or just plain wrong).

Environmental risks (in our sense) are explicitly considered in two areas of this model. Firstly, in the process risk sector as an operations risk along with other elements such as business interruption and product development. An environmental scan is also considered as part of the strategic information for decision making component.

Waring (1998) suggested that environmental along with the following risk areas are likely to be more prominent as we move into the next century: financial risk management, human resources, security, product liability, travel, organisational changes, corporate reputation, territory risks, and occupational violence.

He describes a risk profiling process (often using external specialists) based on interviews, examination of documents and site observations supported by techniques such as brainstorming, differentiating between essentially 'pure risk' areas such as health and safety, fire, security, IT reliability and environmental, where the best that can happen is that nothing happens – this contrasts with 'speculative' risk areas such as HR, political, mergers and acquisitions where outcomes may vary between success and failure.

In addition to environment as a 'pure risk' in its own right it is evident that it could also overlap into many of the other areas described. Corporate reputation would be the most obvious perhaps post - Brent Spar, but looking a little closer could yield scenarios such as an environmentally damaging compound in a product (asbestos?) or territory risks (corporate environmental performance in developing countries?)

An important issue for global companies is the differing perception and attitude towards environmental issues in different cultures. Witness the two contrasting quotes in an article discussing the 1998 environmental incident where an earth tremor disrupted a dam near Seville resulting in the release of at least 5 million cubic metres of industrial effluent, causing widespread damage (Parry, 1998).

> *'The Donana disaster has had a big impact on public opinion in Spain. We also expect that as a result many private companies will take a close look at the environmental area of their risk management policies ...'*

A more sceptical observer noted:

> *'The accident will only have a limited effect on business mentality: European companies will continue to export their risks to areas like Andalusia which have high unemployment and offer them the chance to cut costs and skimp on environmental safety standards.'*

Where commercial activities take place in many locations, the site dependency of environmental risks will be exacerbated by variation in legal controls from state to state. Similarly what may be regarded as unimportant in a home setting may be very important to other cultures (for example, the varying importance attached to contaminated land across Western Europe)

The Association of Corporate Treasurers (Harris-Jones, 1998) identified several possible mechanisms for risk identification including self-assessment, facilitated workshops and company and peer group reviews. Self-assessment could include brainstorming sessions or questionnaires with risk prompts. Workshops, preferably assisted by skilled facilitators and risk specialists, help to increase broad awareness of issues as well as identifying new risks. Historical reviews of operations or assessment of other organisations (in a type of benchmarking exercise) might also be useful.

Since the major benefit of a risk approach is to focus resources where most benefit will be realised, it follows that a relatively small amount of effort expended at the early stage of a venture can result in significant benefits at later stages (or avoidance of possible liabilities). The idea here is that consideration of potential environmental risks as early as possible leads to prevention rather than cure, thus a tiered approach is advocated committing minimal resources initially in screening type exercises. The tiered process is described in terms of levels, a level 1 process representing an initial screening exercise leading on to the progressively more detailed level 2 and level 3 studies.

A phased approach may be made to the assessment of environmental issues associated with any project, joint venture or business activity. In this way the phased assessment associated with property transfer or pre-acquisition may be seen simply as a particular example of this approach. The checklist below is intended to be applicable to almost any type of situation, for example, as part of an internal audit self-appraisal process. Should there be uncertainty about the potential response, recourse may be needed to environmental specialists.

Perhaps the key issue is to look beyond the factory fence, literally and metaphorically. Issues may only become apparent by looking up and down the supply chain. Furthermore, the expectations of customers and others is likely to change over time, so that something that was deemed insignificant last year may now be a major concern.

The checklist in Box 5.1 aims to identify simple downside risks, however by seeking opportunities for resource efficiency, for example, other commercial advantages may be obtained.

Box 5.1 *Strategic Environmental Appraisal Checklist*

Environmental Review Checklist

- **Is the activity subject to legal controls?**
  Legal control mechanisms are most likely to be apparent (eg in terms of the need for prior authorisation) for processes involving hazardous materials, however, there are almost always going to be some applicable laws and regulations. For example, most legislative regimes operate controls on solid waste storage and disposal.
- **Are hazardous materials used in the process or are they incidental to site activities?**
  For example fuel, or natural product storage.
- **Does the activity require the involvement of third parties where environmental concerns are significant?**
  Joint venture partners' reputation, contractors and their standards might all be relevant.
- **Are there environmental issues associated with the use or disposal of the product or service?**
  Electronic equipment and battery powered products might be examples where disposal at end of life is significant.
- **Are there environmental issues associated with the supply of raw materials?**
  Mining and quarrying and tropical hardwoods would be obvious examples.
- **Are significant resources used in the activity?**
  For example high energy use or large quantities of aggregates.
- **Are there historical concerns arising out of the activity?**
  Pre-existing contamination of land acquired might be a significant liability.
- **Is there potential for significant public or stakeholder interest?**
  There is a need to address external perception of risk, as opposed to the view from internal aspects.

In assessing the potential for environmental issues to be championed by external pressure groups, an interesting analysis has been conducted by Winter and Steger (1998).

They developed a simple checklist that can be used to predict whether an issue has the potential to pose serious risks to an organisation, asking questions such as:

- Are arguments against the issue plausible and understandable to the public?
- Is the issue visual, touching, media friendly?
- How strong is the activist groups driving the issue and how isolated is the company?
- How easy is the solution?

They use these and other elements to undertake a retrospective analysis of events such as Brent Spar.

Integrated Project Assessment

A significant number of environmental protocols have been created to deal with projects (as opposed to systems designed to apply to a location, process or organisation). For example, environmental impact assessments were one of the earliest developments in formal environmental management controls. This has been complemented by the project-based nature of many organisations, such as oil exploration companies, working in environmentally sensitive sectors.

A recent publication from the Environmental Analysis Co-operative (EAC,.1999) describes approaches to ensuring that environmental considerations are addressed at all stages of a project, with particular reference to those associated with the process industries. A particular aim was to facilitate the difficult issue of comparing environmental effects with alternative process options.

Two techniques for ranking environmental effects in order of importance are presented – force ranking of pairs and concept ranking. In the former a two-dimensional matrix is prepared with the effects (a, b, c, d, etc) along both axes. Effect a is then compared with effect b and ranked either higher or lower. In the case of higher a score of 1 is entered in the cell, if lower 0. The score for each effect then reflects its relative importance and facilities a ranking.

Concept ranking works by assigning an importance weighting to a category of environmental impact (eg noise or visual impact), then identifies a number of impact parameters within the categories and allocates the overall importance between the parameters. Then for each project option assign a significance factor from 1 (negligible) through to 10 (extremely significant) for each impact parameter. Impact scores are then obtained by multiplying the importance weighting by the significance factor for each parameter. Summing the impact scores for all the parameters for each option then gives total impact scores that can be ranked

In terms of avoiding potential environmental liabilities and ensuring that environmental costs are accounted for, the procedures utilised by the European Bank for Reconstruction and Development (EBRD,1996) are among the most advanced. Part of its founding agreement is to promote environmentally sound and sustainable development in the full range of its activities. The early assessment of issues is viewed as important in assessing and mitigating the financial risks of, for example, increased operating costs to meet legal requirements or historical pollution problems. Since a significant part of the banks investment is directed through financial intermediaries such as regional banks and investment funds, a mechanism is also in place to ensure that they have appropriate environmental controls in place

At an initial review stage in the consideration of new ventures comes the screening process. This is a simple process to categorise projects on the basis of their potential environmental effects and to highlight issues and opportunities. This screening is said to require only 10–20 minutes of effort and determines whether an analysis (category B) or a full environmental assessment (category A) is required or that the potential impacts are very low with no further detailed study required category. The environmental assessment is an intensive examination which might be initiated for a greenfield development or where required by national laws such as the Town and Country Planning (Assessment of Environmental Effects) Regulations 1988 in the UK. It may extend to a strategic environmental assessment that considers a project in the context of an overall programme or economic sector. If an environmental audit is required it is screened as category 1, if not category 0 is applied. The audit requires fewer resources and is often applied to ongoing industrial operations and its scope would be relevant in a broader sense and, in the terms of this chapter, might be called a level 1 or 2 study. It can be considered that the environmental impact assessment is a formalised subactivity within the formal risk management programme. Even when the need for such a regulatory requirement is not triggered, a tiered assessment of environmental effects can provide benefits with no further detailed study required.

The EBRD audit might be considered a level 2 assessment, although the term environmental review or baseline review might be equally valid. The format of the audit report would include an executive summary, an introduction along with a description of the site location and setting. This would consider historical aspects such as the nature of processes and what materials have been used. There would also be assessment of the environmental sensitivity of the site in terms of geology/hydrogeology, nearby population etc. After a description of environmental management systems, policies and procedures in place the emergency and security arrangements are considered along with community interactions and environmental insurance coverage.

The following specific environmental topics are then addressed:

- regulatory compliance summary;
- air emissions;
- raw water and waste water;
- material handling storage and transport;
- hazardous materials management;
- oil filled electrical and hydraulic equipment (PCBs);
- asbestos and related materials;
- waste management and housekeeping;
- noise, vibration and other physical factors;
- radioactive materials;
- contaminated land and groundwater;
- cleaner technology initiatives;
- energy use and conservation; and
- waste reduction.

Product and health and safety issues are also considered. Following appropriate environmental investigation a review memorandum is produced based on the findings of the assessment, audit or

Environmental procedures

Figure 3.1

CONFIDENTIAL

Environmental Review Memorandum

Intermediated financing

Operation Title:

Operation Team:

OGC Lawyer:

1 **Brief description of the operation**

2 **Environmental information reviewed**

3 **Site visits by Bank environmental staff/consultants**

4 **Environmental issues associated with the operation** (e.g. risks and liabilities, regulatory trends and other country-specific issues)
Environmental Risk Rating: [high/medium/low/insufficient information]

5 **Environmental issues associated with the potential portfolio** (e.g. typical compliance problems, impacts, risks and liabilities, benefits and opportunities)
Environmental Risk Rating: [high/medium/low/insufficient information]

6 **Environmental due diligence requirements** (based on FI's existing or proposed environmental procedures, if any)

7 **Operation monitoring and supervision requirements** (e.g. EBRD review of loan/investment proposals and reports, annual environmental reporting by FI)

8 **Outstanding requirements for legal documentation** (e.g. environmental clauses in policy statement, conditionality, requirements for sub-projects, monitoring and reporting covenants)

9 **Other issues** (need for technical cooperation support, public consultation etc.)

10 **Further actions required by Bank staff and the Project Sponsor**

Signature : ______________________ Date : ______________________
Environmental Specialist

Signature : ______________________ Date : ______________________
Environmental Specialist

European Bank for Reconstruction and Development 15

Figure 5.2 *EBRD Review Memorandum*

other investigation combined with results of public consultation and other site visits. The format of an environmental review memorandum for an intermediate financing project is shown as Figure 5.2 (reproduced with permission). Monitoring is undertaken, for example, until the loan is repaid, to provide feedback on the effectiveness of agreed measures. Subsequent sections in the report would deal with legal compliance, conclusions and recommendations for further action.

An alternative (or intermediate) level 2 exercise to a full audit might be a matrix-based risk review of the type undertaken by Dow for new projects (Gowland, 1993). An assembled team assigns elements to a ranking matrix based on severity and probability. Individual matrix cells give an indication of the priority to be attached to the hazard. The matrix and the procedure for completion and assessment of the risk are shown in Figure 5.3 below.

	Severity				
	Very Severe			Low Consequences	
	1	2	3	4	5
High 1	1	2	3	5	7
2	2	4	6	7	8
Probability 3	3	5	7	8	9
4	4	6	8	9	10
Low 5	5	7	9	10	10

Figure 5.3 *Ranking Matrix*

The procedure involves gaining understanding of the processes involved in the project; all the imagined hazards arising should be listed. They are categorised by Severity and Probability from the examples below. They are then plotted on a priority chart as shown above (eg Severity 1 Probability 3 would be priority 3).

The numbers inside the boxes represent ranking for priority (the project team may adjust this) and determine what risk control actions should be taken. Normally there will be no priority 1 items coming from a severe incident occurring once per year, so the list may start with priority 3 items. Usually action should be taken to mitigate the probability or reduce the consequences of hazards with priority 7 or higher.

Severity factors:

1 Fatality off-site or on-site.
Property damage or business interruption > $50million.
Environmental incident like Sandoz/Schweizerhalle. (The Sandoz incident involved a fire at a warehouse in Basle that resulted in the release of toxic materials and subsequent major pollution of the Rhine.)

2 Severe injury on-site or injury (involving hospitalisation) to a member of public off-site.
Environmental incident resulting in clean-up off-site or a fine.
Any event leading to the need to evacuate members of the public.
Property damage or business interruption > $1million < $50million.
Fire or explosion leading to property damage off-site.

3 Any incident resulting in the local public being told to stay indoors, etc.
Environmental incident resulting in a clean-up on-site but which could contaminate groundwater.
Property damage or business interruption > $50,000 < $1million.
Fire that requires site and off-site fire-fighters to control.

4 Minor injury as a result of chemical exposure.
Environmental incident where contamination is confined to site and where recovery is complete in one year. Property damage or business interruption > $1,000 < $50,000.
Fire which is controlled by site personnel.

5 Minor injury from falls, cuts, etc. where treatment is first aid only.
Fire that is controlled by hand held fire extinguishers.
Spill of chemical which causes no operator exposure and which is completely recovered.

Probability factors:

1 Once per year: High
2 Once per 10 years: Moderate
3 Once per 100 years: Medium
4 Once per 1,000 years: Low
5 Once per 1,000,000 years: Very low

Although many sites might consider they would not be capable of a major environmental incident such as the Sandoz event (severity Grade 1), it is likely that any operation is capable of causing a Severity grade 2 incident. This would result in a fine and/or an off-site clean-up operation.

The approach described above could be extended to cover routine emissions. This would require an additional category for both probability and severity factors. The additional probability factor could be something like 'daily or continuous emission associated with routine operation'. For severity a description for the additional category could be worded ‘routine emission within limits set by regulations'.

Integrated Health Safety and Environmental Assessments

The integration of environmental with health, safety and quality management is a theme that has received considerable recent emphasis. Many of the challenges and advantages have been considered in a recent paper by Ramsay (1998). The particular advantage of risk-based approaches as a unifying concept in a world of complexity and uncertainty was contrasted with the development of health and safety and environmental controls as separate structures. This has been because they were seen to be peripheral to the core business management processes. The optimisation of controls in these areas was also viewed to be more difficult when they are managed separately.

A particular issue is the potential 'competition' between the two disciplines. Heath and safety issues having a better established provenance and (in many instances) more formal systems of control and a clearer legal framework in which to operate. Counterbalancing this is the idea that environmental impacts (where they are actively managed within organisations) gain board level interest more easily than health and safety (The CBI noted that health and safety might be perceived as an operational rather than a strategic matter).

Economy of effort can also be obtained by integrating health and safety with the environment; accordingly a means by which this might be achieved is described in this chapter, together with some of the potential release scenarios and consequences for environmental risk assessment. The actual process of identifying and classifying environmental risks can thus be seen as similar to that adopted for health and safety risks. It should be noted that consideration needs to be given to the hazards and potential effects associated with products and processes, not only those under direct control but also taking into account contractors and suppliers.

Bridge (1997) described how the multinational engineering company Vickers extended its Risk Control Programme from addressing property risks to include security, health and safety and environmental issues, starting from an assessment of current environmental status. This involved completion of a 40-page questionnaire by business units world-wide. The information obtained was reviewed by environmental consultants to prioritise sites that required detailed assessments.

There is an established legal framework for risk assessment in health and safety. The Management of Health and Safety at Work Regulations (SI 2051, 1992 amended by SI 2865, 1994) require organisations to implement risk assessment programmes for health and safety which include identification and implementation of preventative measures. The regulations relate to all activities in the workplace and should also address contractors and members of the public. There is currently no legal imperative to consider environmental risks along with health and safety; however benefits might be gained from adopting an integrated approach:

- Many of the materials and processes associated with health and safety risks also raise environmental concerns.

- An integrated approach is cost effective by avoiding duplication.

- Environment becomes a routine part of management control (as has already happened with safety).

A tiered programme that ranges in complexity from simple qualitative assessment of hazards by personnel engaged in the activities to advanced quantitative methods undertaken by experts not only raises general awareness among employees of environmental issues but also ensures that the more time-consuming approaches are only applied where truly necessary.

The initial step in any programme will be to identify hazards, which might arise out of a material itself (eg because of toxicity) or as a result of a process (eg noise or equipment failure giving rise to emissions). It is likely that all health and safety hazards will be considered environmental hazards; however, the reverse might not apply. For example, natural products such as milk or fruit juice might not constitute health and safety hazards, although they can cause significant environmental damage if they escape to watercourses. Similarly, the primary concerns over CFC emissions would be ozone layer depletion rather than health and safety concerns. The decision on whether risks are acceptable and what control measures are needed is subjective for both health and safety and environmental risks and depends on circumstances. It should be noted that the magnitude of environmental liability (unlike health and safety) is likely to vary significantly depending on the site location. It is self evident that individuals who are able to identify release scenarios and likely consequences of a specifically environmental nature need to be included on the team performing the analysis.

These ideas can be viewed in the context of the guidance available from the Health and Safety Executive. For example, a concise guide (HSE, 1994) has been produced for organisations in the commercial, service and light industrial sectors.

The five step process involves:

1 Looking for the hazards – conducting a walkover and making observations and talking to employees and their representatives.

2 Deciding who might be harmed – including people who might not be in the work place all the time (eg contractors) and the general public.

3 Evaluate the risks.

4 Record the findings.

5 Periodically review and review the assessment.

HSE Level 1 Screening

In the level 1 screening appraisal a line supervisor (or skilled operatives) carries out a preliminary identification of risks and an assessment of whether they are significant. There is likely to be some familiarity with what might constitute health and safety risks; however, a short briefing or written prompts might be useful for the environmental aspects. Examples might be:

- major use of resources (energy, water, materials etc);
- use of 'black list' materials (lead, cadmium, mercury etc);
- emissions to the atmosphere (organic solvents etc);
- discharges to water (routine discharge or accidental loss); and
- solid waste production (hazardous waste).

It might involve a supervisor walking around the worksite, noting and logging hazards. It should be emphasised that unusual circumstances should be considered (eg when key staff are on holiday) as accidents might depend on a number of situations occurring simultaneously. Placing Y (Yes) or N (No) next to the identified hazard will indicate whether further consideration would be required. Should the level 1 approach indicate a significant hazard, the level 2 assessment is triggered.

HSE Level 2 Assessment

In the level 2 assessment a team of 2–4 persons estimates probability and severity for each of the hazards identified. The group could consider risks by:

- process;
- activity; and
- area.

They should try and consider:

- assembly of x, maintenance of y, etc;
- use of hand tools, etc;
- work in section a, work in department b, etc;
- all aspects of any operation and worst possible outcomes arising;
- any personnel particularly at risk, eg visitors, contractors or new employees;
- whether existing controls are satisfactory;
- probability of occurrence and severity of outcome of the event, assigning numerical values to each; and
- special environmental considerations, eg whether the operation is taking place in a sensitive environmental setting.

Whenever a work system (or location) is changed, the exercise should be repeated. Should there be a serious incident the assessment should also be checked and revised as appropriate. For complex operations it might be worth breaking the assessment down into elements:

- installation;
- start up;
- normal operation;
- abnormal operation;
- maintenance;
- shut down; and
- decommissioning.

The product of the probability and severity is calculated and if this is greater than the trigger value, a higher level of assessment is initiated unless controls can be easily added to make the risk acceptable.

Controls could include:

- elimination – stopping the activity;
- substitution – using alternative materials;

- implementation of written procedures;
- supervision of activity;
- use of protection equipment; and
- training.

The team could utilise guidance on the nature of environmental incidents and associated initiating events. In particular the importance of considering impacts to surface and groundwater needs to be accommodated, along with potential liabilities associated with contaminated land and hazardous waste disposal.

An example of an assessment of an individual activity for frequency and consequences could be the loading and unloading of fuel oil shown in Table 5.1. It should be noted that incidents involving releases to surface water are much more visible than those to groundwater and hence much more likely to be identified (and remedied quickly). Releases to groundwater can continue undetected for long periods and require much more effort to remedy contamination.

Frequency factors could be assigned based on the number of events per year (eg 50 equates to a weekly delivery). The consequence factor could be based on whether a significant spill would be retained in the immediate vicinity (ie by a bund or other containment around the area) which would give a consequence rating of 1; it could be retained on the site by an interceptor or holding tank to give a consequence factor of 2, or escape off-site to give a value of 3. Consideration should also be given to the number of facilities on-site since a single unit might not be of concern but a group of facilities collectively could represent undesirable environmental risks.

Table 5.1 *Activity Assessment*

Activity	*Frequency Factor*	*Consequences*	*Risk Rating*
Delivering Fuel Oil	50	3	150

HSE Level 3 Assessment

Subsequent higher level assessments could include a formal HAZID (Hazard Identification) in which a team visualises ways in which a system, equipment or tasks can lead to risks. The imagination of the team needs to be stimulated in a systematic yet creative fashion. In order to achieve this end, which is equivalent to a Hazard and Operability (HAZOP) study, the following should be realised:

- A full description of the operation or task.
- A systematic examination of each part of the system to assess where deviations can occur from the intended design.

Guide words are used in the study in order to ensure that all conceivable design deviations are considered. Each deviation is then considered in terms of how it may be caused and what the consequence of its occurrence would be. Identified events that are both conceivable and have significance are detailed for remedial action. Should the solution be obvious, the control measure can be implemented directly. However, it may be the case that further detailed study is required involving quantitative estimates. The next level of work is then requested.

In terms of additional elements required to address environmental risks, it is unlikely that further initiating events will need to be considered, ie the probability factor is likely to remain the same. This would not be the case with the consequence element as new scenarios should be considered. The guide words must therefore reflect the additional pathways and receptors for environmental risks such as:

- via drains or surface run off to surface waters;
- via ground to surface water or groundwater; and
- via air to sensitive ecosystems.

The consequences of relatively minor releases over a long period could equate with a major release over a short time, hence the ease of detection of a release (eg above ground vs underground) is also a factor. Should site-specific information be available on geology, hydrogeology etc, this can certainly influence the consequence and should be used. In considering control aspects such as fire protection, environmental aspects should also be included as contaminated firewater can give rise to significant problems. Should it be decided that the operations pose such a significant risk and that this may not be readily controlled, then a still higher level of investigation such as fault tree analysis is warranted.

Fault tree analysis considers the probability of events leading up to failure, by which means the most likely cause of failure can be identified and controls instigated to prevent the top event occurring. Drawing the tree starts with the top event and moves down to consider all the possible events that might lead up to the top event. Probability values are then assigned to each of the events in the tree, and the overall probability of the top event obtained by combining the probabilities up the tree.

Since this method deals only with the probability of an event occurring it can be seen that no additional elements might be required when considering environmental risks. This applies on the assumption that environmental hazards and consequences have been identified at an earlier stage. A pure safety fault tree analysis might thus be considered to be adequate in terms of scope and expertise required, even when the primary motivation might relate to prevention of a previously identified environmental incident.

Other key factors in the avoidance of accidents are as follows: firstly that plant or process changes are reviewed and approved by specialists before implementation, and secondly, vigorous control is needed to cover the movements and activities of personnel from outside organisations such as contractors and visitors. These people may be less well supervised and controlled than in-house employees. It is also important to have a culture which encourages the reporting of 'near misses' as well as actual incidents. Study of these incidents can prove extremely helpful in the avoidance of future incidents. Clearly an atmosphere of trust is required to ensure that the system of working is studied rather than apportioning blame to the individual employee.

Environmental Management Standards

Many organisations in the process industries have been developing and operating their own environmental management systems and conducting environmental audits since at least the 1970s, particularly in the US. These were initially concerned with assuring compliance with relevant legislation. It was from these audits that environmental management system concepts were developed.

During 1992 The British Standards Institution published BS 7750 (reissued in 1994) which was the worlds first environmental management standard. The European Commission Regulation 1836/93 Allowing Voluntary Participation by Companies in the Industrial Sector in a Community Eco-management and Audit Scheme (EMAS) became operational in member states in April 1995. This Regulation drew heavily on the pioneering work of BS 7750.

More recently, an international environmental management system standard ISO 14001 (BS EN ISO 14001 is the full title of the UK version) was published in 1996. The US and Pacific Rim countries were actively involved in its development, partly because of concerns that other standards such as EMAS could be used as barriers to trade. As a result of international influence ISO 14001 is somewhat less prescriptive, but fundamentally similar to, BS 7750, which it superseded.

ISO 14001 draws upon management system standards such as ISO 9000 that provide a general framework for internal control and can also be assessed and certified by a third party. Although originally intended for application to manufacturing industry, it has been written so that it can be applied to almost any type of organisation.

The standard requires a written policy statement validated by a senior member of the organisation, which demonstrates a commitment to environmental improvement and provides overall direction. This policy would be supported by procedures (work instructions) that describe how jobs are carried out in an environmentally responsible manner.

Any business registered to ISO 14001 must therefore commit itself to continually improving its own environmental performance, with legal compliance regarded as a minimum starting point. There is considerable emphasis on identifying the relevant environmental legislation and on drawing up objectives and targets for performance improvement. The performance targets should reflect the most important environmental aspects associated with operations, accordingly a mechanism is needed to identify the significant aspects. The targets and objectives for improvement are set by the organisation itself; however, reference can be made to other guidance such as codes of practice. The progress made is then monitored, for example, by internal checks, measurements and audits.

Some organisations might wish to make progress towards environmental improvement in a systematic way, but without being committed to obtaining external certification. A realistic approach in these circumstances could be to undertake tasks, consistent with ISO 14001 but not pursue third party certification.

The ISO 14001 specification includes the following elements:

a) environmental policy;

b) planning, which includes establishing procedures for identifying environmental aspects and complying with legislation;

c) implementing and operating an environmental management system, including identifying responsibilities, training and emergency preparedness;

d) checking and corrective action, which includes monitoring, action in the case of non-compliance, record keeping and auditing; and

e) environmental management reviews.

The basic concepts and thinking behind the various elements are relatively straightforward:

- A preparatory review or other available information (eg from previous audits) identifies the range of environmental effects associated with the operation.

- A commitment from the most senior level, developed and explained in the organisation's environmental policy, to manage and control the aspects.

- There needs to be a process to determine and document what legal obligations exist and to decide what are the important environmental impacts associated with the organisation.

- Objectives are then set to control and improve performance on these important aspects.

- Progress towards achieving the objectives is monitored using suitable measurements, internal audits, and management reviews.

EMAS aims to encourage public reporting on environmental performance. An environmental statement (report) must be prepared for each site at the same frequency as the internal audit and be based on the findings of the audit (or the initial environmental review). An accredited (external) environmental verifier must validate each statement. The requirement for external verification and production of a public report represents one of the major differences between the EMAS system and ISO 14001. As with ISO 14001, participation in EMAS is voluntary. The scheme has been limited to companies involved in specific industrial activities at individual production sites and has not therefore been applied at present to a wide range of non-industrial organisations. The regulation has been reviewed and it is likely that it will apply more broadly in future.

A report (ENDS, 1997c) has indicated a significant variations in cost per employee of implementing and maintaining certified environmental management systems. The cost of implementation (including costs such as diverted annual resources, external consultancy and certification) ranged from less than £100 to more than £250 per employee. The average cost of maintaining a system was £37, ranging from just under £25 to more than £350 per employee per year. The average external cost of obtaining certification was given as £10,000 where consultancies were used and £2,000 as the average annual external cost for maintaining the EMS.

The approach described in environmental management standards such as ISO 14001 does not contain any great surprises for a competent manager. Reduced to their essence they may be regarded as common sense measures that seek to identify relevant environmental impacts and then put measures in place to control these impacts. A variety of approaches can therefore be adopted in the use of the standards. They can be used as useful reference points to assess existing controls or at the other extreme involve a documented system that has achieved successful registration following assessment by an accredited third party organisation. The latter approach will clearly involve a greater commitment of resources and the decision to pursue this path will be principally commercial.

ISO 14004 is of particular relevance as it provides general guidance on environmental management and its successful implementation (without necessarily obtaining third party certification). An advantage of third party certification is that it demonstrates to customers and other interested parties that appropriate controls are in place. In order to gain third party certification, there must be records that the outside body can check to ensure that what the organisation says it does is actually happening in practice.

Since adoption of the standards is voluntary there clearly must be a commercial imperative to adopt them. In heavily regulated sectors the assurance of compliance or liability minimisation may suffice, however, other factors may play a role. One of the key elements could be a desire to maximise efficiency in the use of resources such as energy or waste. Corresponding cost savings will also be gained. Demonstration of environmental responsibility to third parties such as investors and pressure groups could also be relevant. Registration may be a pre-requisite for gaining entry to tender lists, for example, Rover Group insisted its major suppliers were ISO 14001 registered. It could also be used to demonstrate environmental probity in mitigation should an incident arise. In a reported case (Environment Business, 1997) 'EMS system cuts fine by £70,000' a company received a relatively low fine after pleading guilty to environmental offences with environmental management controls being cited as a mitigating factor. There is also ongoing debate as to whether regulators should recognise environmental management standards, for example, reducing inspection frequencies.

Environmental risk assessment is also relevant in recent environmental management system standards such as ISO 14001 in the context of evaluating environmental effects arising out of emergency conditions where the probability of an adverse event occurring and the severity of the environmental effects arising out of the incident both need to be considered. This assessment would then form the basis for establishing objectives, controlling risks and establishing environmental components of emergency plans. This approach can be seen as extending the traditional scope of health and safety studies by considering a broader range of materials to be hazardous (eg stratospheric ozone depleting substances such as CFCs) and considering ecosystem targets in addition to humans. It is, therefore, possible to integrate environmental issues within other health and safety requirements such as the self assessed risk of routine operations. Although a minor spill of several litres of oil for example, could in no way be considered a major incident, it is inherently more likely to occur. A simple subjective risk assessment offers the potential to reduce overall environmental effects by minimising the probability and consequence of such relatively minor events.

There remains scepticism that the application of the standards is universally successful. Waring and Glendon (1998) observe that:

> *'The high failure rate for full absorption of these add-on management systems (such as ISO 14001) at least partly reflects the different cultures or world-views respectively (a) required for their successful adoption and (b) represented within organisations as collections of sub-systems with divergent cultures and interests.'*

Future Developments in Environmental Risk Management

In order to achieve the broader interpretation of environment as a mainstream business risk it is worthwhile considering some of the issues that might need to be addressed in future. Foremost among these is ensuring related financial information is available and utilised. Epstein (1996), in a study of environmental management in 100 leading corporations, indicated disappointments including:

- poor co-operation between environmental and other corporate functions;
- a common focus on compliance rather than an ISO type planning approach; and
- inadequate systems for identifying and tracking environmental costs.

The greater adoption of ISO 14001 type systems could provide the means of addressing these problems. Nonetheless, environmental management system implementation is often sold with the promise of minimal disruption to business, which might run counter to the integration ideal (except in the sense of existing similar systems). This could lead to a lack of cost-related information being gathered, which of course is the fundamental basis of business decision making.

Woellner (1998) recognised the disconnection between decision makers and the (mostly technical) people to whom responsibility for environmental issues had been delegated. The range, novelty and complexity of environmental risks seeming to exacerbate this process. He considered an approach within an ISO 14001 framework would put environmental management into the realm of strategic decision making for the following reasons:

- it focuses on management systems;
- it moves away from a command and control approach looking at 'end of pipe' solutions;
- it encourages root causes analysis; and
- it focuses on ongoing risk control and improvement.

Environmental Management as a Financial Issue

Fines and damages associated with prosecutions for environmental offences are often cited as financial costs. Although these may be high profile they may be a poor indicator of overall costs. Almost 20,000 substantial cases of water pollution were dealt with by the Environment Agency in 1997. By way of comparison, 169 successful prosecutions were gained (of course there is a time delay in the court cases so the prosecutions reflect some 1996 incidents and some 1997 incidents had not been resolved).

Prior to the recent £4million fine for the Milford Haven Port Authority (Environment Agency, 1998), the largest fine under the Environmental Protection Act and the Water Resources Act was £300,000 following a major spill of chloroform. The highest fine under the Water Resources Act alone was £12,000 with a range of £500 – £12,000.

There is no readily available database by which the costs associated with environmental incidents may be readily assessed. The probable cost of the most frequent type of accidental environmental pollution has been reported as zero (since the consequences are not severe or go unnoticed) (Smets, 1988).

A recent study (Davies, 1999) has looked at the cost associated with risk by investigating a number of accidents. It was found that even substantial fines can be dwarfed by other costs incurred. In 75 per cent of cases where environmental clean up was needed the associated costs were greater than that of the legal costs and fines combined. Similarly, the cost of third party compensation and the implementation of improvement measures post event could also be important. A great variability in the costs from accident to accident was also noted.

Another important contribution to understanding these elements has been published by the Health and Safety Executive (HSE, 1997) which looked at accidents of a routine nature (rather than those highly visible events associated with major catastrophes). Using a broad definition of accident that included ill-health injury, damage to property and the environment, five case studies were to look at all losses over a study period. The results were as shown in Table 5.2.

Table 5.2 *HSE Financial Losses*

	Annualised Loss	*Representing*
Construction site	£700,000	8.5 % of tender price
Creamery	£975,000	1.4 % of operating costs
Transport company	£196,000	1.8 % of operating costs
Oil platform	£3,763,000	14.2 % of potential output
Hospital	£397,000	5 % of running costs

Both financial and opportunity costs were considered, financial being the costs incurred in material and labour to restore a situation. Opportunity costs were considered as arising from lost opportunities including the cost of people not being able to do their job.

Our scope of interest is restricted to environmental risks. Although there is no clear cut off between what constitutes environmental, some of the scenarios described here have a stronger environmental component. At the creamery over £8,000 of losses were caused when bag filters became over pressurised and milk powder escaped. Money was spent on cleaning the construction site following a leak of sewage and nearly £7,000 costs resulted from an oil spillage at an NHS hospital when a tanker driver attempted to fill an already full storage tank. The environment is likely to contribute a lower proportion overall as it is often linked to potential for off-site impact, although it is very difficult to quantify the division. It should be remembered that these scenarios do not constitute the universe of environmental risks as elements such as supply chain and planning consent issues and reputational matters were not included. Furthermore, some environmental incidents (eg leaks to ground) may have no apparent short-term costs but very significant long-term costs through contamination of groundwater, for example. There were no catastrophic events or prosecutions during the studies further supporting the idea that legal fines and damages are an extremely poor measure of overall costs.

In follow up work undertaken by the HSE a more restricted definition of accident was used that excluded environmental (and opportunity) costs. The significantly lower costs obtained are consistent with the proposition that environmental costs form a significant proportion.

Upgrade or remediation costs might also be important. A 1993 survey (Roche, 1996) of around 1,000 landfill sites showed that 18 per cent had suffered a significant pollution incident or failure that required substantial remedial treatment. Landfill gas migration was the most frequent form of problem affecting around 1 in 10 sites. Surface water pollution incidents affected more than 1 in 20 and groundwater pollution more than 1 in 40. A second phase of the survey showed that the remediation costs for groundwater pollution were greater than £100,000 and in some cases in excess of £1million. Landfill remediation schemes typically cost in the order of £100,000 and sometimes more than £500,000. Surface water pollution incidents are generally less costly to remediate typically in the range of £5-15,000. Interestingly more than one-third of the sites were considered to be of the 'containment' type with 5 per cent commenced after 1990.

Criticism has often been levelled at the financial community that, given their significant potential influence on corporate activity, it does not pay sufficient regard to environmental issues. From the perspective of the investment analyst, however, the picture can be somewhat different. Firstly, there is a common view that as an 'ethical' issue, the environment is principally of interest to the ethical investment community. Secondly, and perhaps most importantly, the argument that improved environmental performance leads to increasing shareholder value has still to be proved (implicit in this is the assertion that environmental costs are material). Public environmental reporting may be satisfying the requirements of many audiences, but in overall terms it has had limited utility for the investment community. Firstly, there is little data on financial consequences, and, more importantly, it is very difficult (within a sector) to compare one company with another.

Establishing a link between environmental and financial performance would ensure that the environment could be integrated within mainstream investment decision making. Edwards (1998) discussed research in this area that included US studies suggesting that companies are not paying a financial penalty for improved environmental performance. Research was undertaken based on fifty listed UK companies assessed over the period 1992–96. These companies had been assessed by a specialist investment house as being environmentally best in their class based on factors such as the monitoring of environmental impact and environmental management controls. The financial performance of these companies (in terms of Return on Capital Employed (ROCE) and Return on Equity (ROE)) was compared initially with the average performance of similar non-green companies and then the best financial performance from the non-green sample. Results indicated a generally better financial performance from the green companies than the average over a range of industrial sectors and also that they did not appear to be suffering any performance disadvantage when compared with the best non-green companies.

One approach that is attempting to address some of the issues relating to the investment community is an environmental risk rating initiative has been promoted by the Safety and Environmental Risk Management (SERM) rating agency (IEM, 1998). This scheme involves the calculation of a rating analogous to credit ratings, on a 27 point scale from AAA+ to C-. The rating process involves:

1. Identification and costing of safety and environmental hazard scenarios for the organisation being rated (based on publicly available information, interviews and site visits). Insurance cover in place is then offset against these direct costs, covering claims such as fires and environmental clean up.

2. Indirect costs associated with the scenarios (lost sales due to protest, loss of lender confidence, reduced employee morale etc) are then calculated.

3. The overall inherent cost of risk (capitalised residual risk) is then calculated by combining the costs and probabilities of the scenarios. The probability of each scenario is adjusted to reflect the effectiveness of management controls.

4. The capitalised residual risk is then compared against the company's market capitalisation to determine the rating.

Eastern Group is the first major company to have commissioned a rating (AA was achieved). In order for this initiative to be successful it clearly needs to demonstrate its worth to the financial analysts and the rated companies. There will need to be sufficient numbers of companies rated to enable benchmarking.

Recent research (Figge, 1998) has postulated some important features in relation to environmentally-linked investment, which could influence investment strategies. Diversification is seen as a key tool in reducing overall risk in investment portfolio management. However, this only applies where the investments are effectively independent (which could be in the traditional sense or in an economic sense). Losses on one stock or investment period are accordingly balanced by gains in other stock or periods. However, this may not be the case with some environmental risks. Important environmental risks are just appearing without a loss history to facilitate estimates of harm, therefore a greater uncertainty is incurred. The future role of fuel cells and solar energy is quoted, whereby a future fall in price could pose (investment) threats to traditional power generation. Environmental risks can have a strong systematic element and operate over extended time periods, hence removing the opportunity for diversification. The example quoted by Figge was that of climate change, where control measures (such as an energy tax or levy on CO_2 emissions) would have a broad cross-sectoral (ie-systematic) impact that would apply for many years. Ozone layer depletion and hormone disrupting chemicals were two further examples noted.

CHAPTER 6
DECISION MAKING WITH UNCERTAINTY

Much of the discussion in this publication focuses on the assessment of risk, methodologies such as RAGS for Superfund being used where adequate information is available. However, even when quantitative estimates have been obtained, these do not necessarily facilitate management of the risk issues arising. This chapter considers the output of risk assessments and how it might be used to facilitate decision making.

It should be clearly established that uncertainty and risk are not synonymous. This distinction was highlighted by Frank Knight early in the 20th century:

> '*Uncertainty must be taken in a sense radically distinct from the familiar notion of Risk, from which it has never been properly separated. It will appear that a measurable uncertainty, or risk proper... is so far different from an unmeasurable one that it is not in effect an uncertainty at all'* (quoted in Bernstein, 1996).

Thus, in some circumstances, there may not even be sufficient information to establish a risk distribution to facilitate control decisions. In this situation of high uncertainty, less quantitative techniques might be most appropriate. Fuzzy logic, for example, seems to offer considerable potential to deal with complex issues characterised by uncertainty.

When they can be calculated, probabilistic estimates of risk might be compared against some type of benchmark (eg 10^{-6} incremental risk), or different options compared. Regret factors might be incorporated to accommodate the risk-taking propensity of the decision-maker. There has, however, been increasing recognition that risk distributions might better support the decision making process as opposed to single point values.

Problems with the use of point values in quantitative risk assessments were highlighted in a 1992 paper (Thompson, 1992). Firstly, by use of a combination of moderate, reasonable maximum or worst case assumptions it is difficult to assess the degree of conservatism in any assessment. Secondly, by choosing single values to account for the uncertainties in all the variables the assessment may consider scenarios that will rarely, if ever, happen.

Finley and Paustenbach (1994) undertook a detailed comparison on the use of 'point' estimates and 'probabilistic' methods, using Monte Carlo analysis of probability density functions. They noted that the point approach effectively lumped together uncertainty as it related to either 'lack of knowledge or variability' (additional data gathering can reduce the lack of knowledge uncertainty but not the inherent heterogeneity). They found that the 95-percentile value derived from a Monte Carlo based approach was 3–5 fold less than the corresponding reasonable maximal exposure (RME) estimate. For scenarios where indirect pathways were involved (eg eating beef from cows that graze on contaminated soils) the probabilistic 95-percentile value is often an order of magnitude lower than the RME.

The Monte Carlo simulation process involves assignment of a probability distribution to each variable. Single point value estimates can still be used within the framework, if desired, in conjunction with distributions. A single value is then chosen at random for each variable on the basis of its probability distribution. The single values are then combined to give a single point result. This is one iteration and the process is repeated, typically for several hundred or thousand further cycles to build a distribution of results. While there are clear advantages over using single estimates it should be noted that there are difficulties in using this approach. Firstly, decisions must be made on the nature of the distributions to be used and the final output is a distribution that does not necessarily offer a clear-cut rule for decision making. A number of PC based software packages using standard spreadsheets are now available to perform Monte Carlo simulations such as @risk ™ and CrystalBall ™. Interestingly, their primary market appears to be in the area of project finance and capital investment appraisal. A report from the USEPA (1997) has highlighted some of the key principles involved in the greater application of Monte Carlo analysis including sensitivity analysis to enable identification of the key variables. It also recommends that point estimates of risk are made for comparison purposes when Monte Carlo methods are employed.

A US study by Merkl and Robinson (1997) has considered environmental decision making and the particular problems arising from the legal and technical complexities associated with environmental issues, exacerbated by emotional and public relation concerns.

In a hypothetical case study of an oil company with a portfolio of approximately 100 sites they indicated that reductions of 20 to 40 per cent in liability costs are feasible when a risk management approach is adopted. Their recommended approach contrasts with the idea that greatest efforts are focused onto sites with the greatest liabilities (driven by the legal compliance and PR concerns mentioned above). These sites are often poor candidates for significant investment and greater effectiveness could be achieved by considering two criteria – the possibility that a liability might increase in scope and cost and the availability of remediation solutions. For example it would be better to spend resources on a $500,000 problem that is worsening rapidly rather than on a $10million site that is effectively stable.

An initial step was to categorise the sites on the basis of cost escalation and solution availability criteria:

- *Immediate Remediation*: where the potential for cost escalation is high and effective clean ups are readily available, rapid action is mandated. In practice this often does not happen because there may be little current regulatory or stakeholder pressure to initiate action.

- *Immediate Investigation*: sites where there is high cost escalation potential but no ready solutions, a more detailed investigation and/or risk assessment is favoured. For example, could risk-based site-specific clean-up standards be generated to the satisfaction of the regulators that reflect particular end uses for the site?

- *Opportunistic Remediation*: several sites have available remedial solutions and are relatively stable. This stability indicates that other risk-reduction investments might well be preferred in the short term.

- *Monitoring*: at stable, non-threatening sites where efficient clean-up solutions are not obvious, monitoring for potential escalation is favoured. This category is considered to contain a significant number of the overall portfolio of sites that traditionally will have been viewed as problematic and costly. Investment will effectively be moved away from these sites to those that might be deteriorating. Of course, regulatory or other third party involvement will remain very important and will have a major influence on the investment profile.

Epstein (1996) cited examples from the New York State Electric and Gas Corporation where a risk-based priority setting approach is complemented by net present value analysis to demonstrate estimated savings. Another utility sector example was cited in relation to underground storage tank replacement programmes where the costs of risk mitigation against the reductions in financial risk from potential leaks was assessed for different tank designs and leak detection mechanisms. It was found that early removal of some steel tanks where leak detection was not practicable was the favoured economic option with a saving of over $1million over the scenario of waiting until action was compelled by legal requirements.

Remediation Decision Analysis

The following example, illustrated in Figure 6.1, represents a simple decision analysis applied to remediation selection based on the criterion of minimum expected financial cost. The total expected financial cost for each scenario is obtained by combining the costs of different remedial alternatives, with the probability of different scenarios. The approach is applied to the situation where a remedial action is required in relation to a landfill operation where off-site migration of leachate threatens nearby residential areas. The object of the decision-maker will be to choose the least-cost alternative that will remove or mitigate the environmental threat. Professional judgement by experts will be utilised with uncertainty estimates where data is unavailable or where the application of a remedial measure is subject to an uncertain outcome (this is self-evident since the true outcome of remedial actions can only be known in the future).

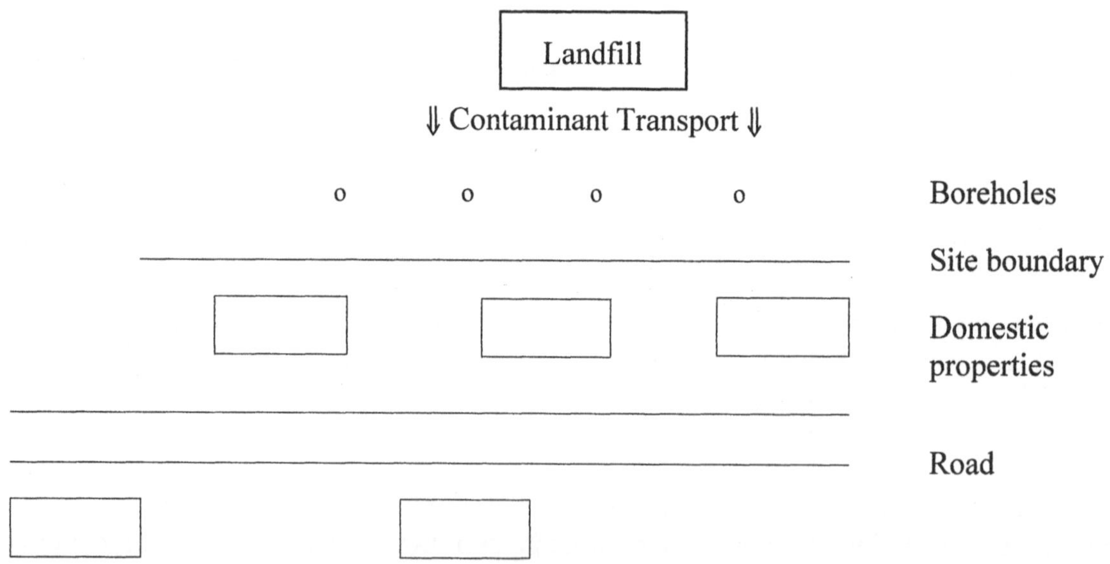

Figure 6.1 *Landfill Leachate Migration*

A number of boreholes have been sunk near the boundary of the landfill property and been found to contain elevated levels of certain contaminants.

Professional judgement is then employed to:

- estimate the probability that there is a significant environmental threat to the nearby dwellings;
- potential remedial actions that could be employed and estimate their likely cost; and
- assess the probability that the remedial action employed will be successful in meeting acceptable standards.

Four scenarios have been identified for consideration:

- no action;
- installation of a passive collection system;
- installation of a pumped collection system; and
- positioning a containment wall between the site and the houses.

Using a scale of 0–1.0 to represent uncertain to certain, a probability of 0.9 was assigned by the technical experts to the probability that the leachate would pose an environmental threat, on the basis of the analysis of the water from the boreholes (ie a 90 per cent chance that it would be a problem).

The cost of each of the remedial actions and the likelihood that protection would be gained are given in Table 6.1. Should all remedial actions fail, the predicted consequence is that liabilities of £100,000 will be incurred for clean-up costs and compensation.

Table 6.1 *Remedial Action Options*

Remedial Action Options			
Action Number	*Remedial Action*	*Implementation Cost*	*Probability of Protection*
1	No Action	£ 0	0.05
2	Passive System	£ 10,000	0.40
3	Pumped System	£ 20,000	0.65
4	Containment Wall	£ 50,000	0.95

The remedial alternatives selected for assessment are 5 in total; 1 to 4 as above, with another involving implementing 3 in the first instance, then, if failure ensues, number 4.

The costs and likelihood of protection can then be combined to calculate an expected cost for each alternative. A decision tree is constructed as in Figure 6.2, with each branch representing a sequence of events. The decision process runs from left to right, with branching indicating where the outcome may be successful or fail to resolve the situation. Calculating the expected cost is achieved by moving along the tree from right to left, combining the costs at various stages along with the probabilities. The first branching after the selection of the remediation represents the probability that the contamination represents an environmental threat.

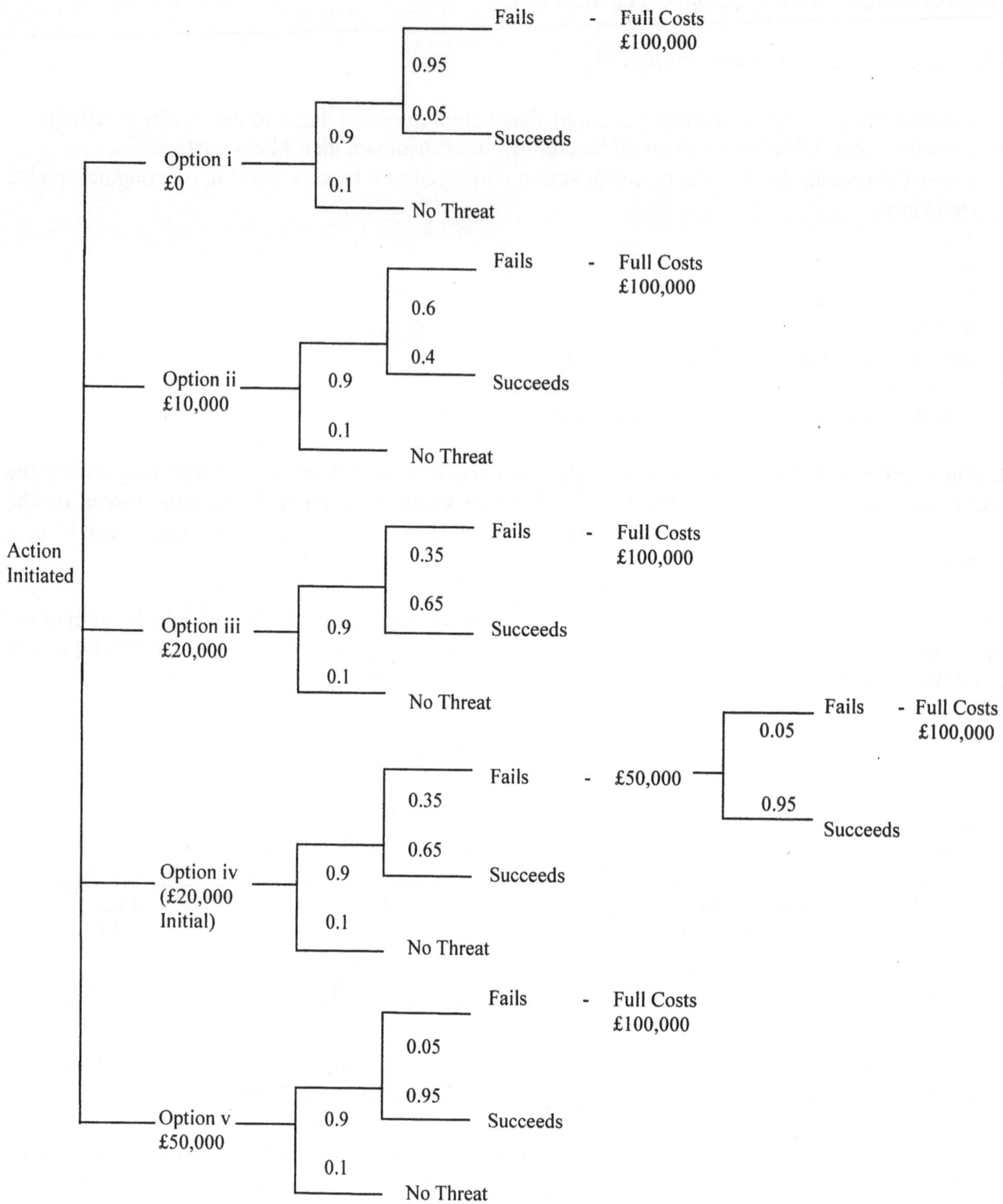

Figure 6.2 *Remediation Decision Tree*

Option costs are calculated as follows:

Option i: No action

Total Cost = £0 (implementation cost) + (£100,000 x 0.95 x 0.9)
= £85,000

Option ii: Passive system

Total Cost = £10,000 (implementation cost) + (£100,000 x 0.60 x 0.9)
= £64,000

Option iii: Pumped system

Total Cost = £20,000 (implementation cost) + (£100,000 x 0.35 x 0.9)
= £51,500

Option iv: Pumped system followed by containment wall (if pump fails)

Total Cost = £20,000 + ([£100,000 x 0.05 + £50,000] x 0.35 x 0.9)
= £37,325

Option v: Containment wall

Total Cost = £50,000 + (£100,000 x 0.05 x 0.9)
= £54,500

Several things should be borne in mind when making decisions on the remediation option. The probability of there being a threat clearly influences the no-action alternative; once the probability becomes reasonably high, adopting this approach clearly becomes more difficult. Options that cost more and offer a lower degree of protection can clearly be excluded at a very early stage. Option iv is preferred on the basis of the minimum expected costs criterion. The illustration used here is obviously fairly simple; however, in real life situations other factors would come into play – for example, additional analytical data could be gathered at an early stage (at additional cost of course), which might affect the overall probability and estimate. It is also important to recognise that the final expected cost figure used for comparison will not be the actual cost but a probability weighted estimate.

Regret and Decision Making

The idea that decision making under uncertainty involves more than the calculation of probabilities to obtain the most favourable outcome, has been considered in some detail by Dembo and Freeman (1998). Talking principally from a viewpoint on investment decisions, they investigate the extent to which the risk propensity of the decision-maker can influence the most appropriate choice. Although infrequently considered, their perspective is intuitively sensible (the extent to which an individual factors in the possible adverse outcome reflects whether they are risk averse or a risk taker). A simple example demonstrates this thinking. Many people are willing to make a small wager in order to participate in a lottery (because the loss of the wager is not seen as particularly onerous), but if the cost of the tickets reflected most of the individuals property the odds they would want before participating would need to be much lower to be attractive (as the regret involved in losing is much higher).

Daniel Bernoulli highlighted this as '*utility resulting from any small increase in wealth will be inversely proportionate to the quantity of goods previously held*', which Bernstein describes as one of the great intellectual leaps in the history of ideas (Bernstein, 1996). Dembo and Freeman illustrate the idea of decision making while considering regret in the context of a BSE type situation. For our purposes an example relating to the purchase of property that might be contaminated is perhaps more relevant.

We wish to purchase a portfolio of ten properties that may or may not be contaminated. Prior to the sale we can pursue various levels of intervention in terms of identifying information relating to the site. Under a low intervention scenario we decide to buy the sites 'blind', which of course entails no additional cost. In the medium scenario we undertake Phase 1 surveys at each of the sites, which means we have a higher probability of avoiding contamination (or dealing with it appropriately as part of the contract). The up-front cost for this option is 10 x 2,000 = £20,000. Should we be particularly concerned about this issue, we could initiate more detailed site investigations costing £10,000 each, which provides us with more comfort in avoiding contamination of the higher initial cost of £100,000.

The outcome of the transaction in terms of financial cost (cost of investigation + cost to remedy contamination) are shown in Table 6.2 for four different scenarios. In scenario one no significant contamination has been encountered. This contrasts with scenario four where the low intervention decision has resulted in the acquisition of three contaminated sites (one seriously damaged requiring clean-up costing £750,000, one costing £100,000 and another £50,000). As a result of the greater investigation, the medium- and high-intervention approaches have incurred smaller clean-up costs of £170,000 and £100,000 respectively.

From Table 6.2 we can start to rationalise decision making behaviour. Should someone feel lucky they may adopt 'minimin' (minimum of minima) cost approach, which means do nothing up front (and hope the sites are clean). Alternatively, a very cautious individual might chose the 'minimax' minimum of the maxima. We would then adopt a high intervention approach, as the maximum cost here is the lowest maximum (at £200,000) of the three action options.

Following Dembo and Freeman's approach, the next step would be to include 'regret' in the calculations. This would consider the best outcome from each scenario and regard the other options difference in cost as a measure of the regret. For example, in scenario two, low, medium and high intervention have all resulted in the acquisition of a single lightly contaminated site (remediation cost £50,000). The regret for medium intervention is therefore £70,000 -£50,000 = £20,000, and for high £150,000 - 50,000 = £100,000.

Table 6.3 shows the costs incorporating regret. Minimax rules (or whatever) can be applied again. In this instance we would choose the medium intervention with a maximum of £70,000. This happens because the up-front cost of the high-intervention approach means significant regret occurs if little contamination is realised. The reason for not choosing the no-action approach is even clearer as in the worst case it results in the acquisition of seriously contaminated sites. Dembo and Freeman develop the idea further to arrive at a risk aversion value (λ) to describe behavioural differences.

Table 6.2 *Property Acquisition – Costs*

		Scenarios			
		1	**2**	**3**	**4**
Intervention Level	*Low*	0	50	150	900
	Medium	20	70	120	270
	High	100	150	150	200

Table 6.3 *Property Acquisition – Costs with Regret*

		Scenarios			
		1	**2**	**3**	**4**
Intervention Level	*Low*	0	0	30	700
	Medium	20	20	0	70
	High	100	100	30	0

Fuzzy Logic

A discipline that would seem to have considerable potential for application to environmental decision making is that of fuzzy logic. Although most visible through its use in commodity products such as camcorders and washing machines, it has characteristics that make it particularly suited to environmental issues. This is because it offers a framework to deal with the great complexity and non-linearity of systems involving living organisms. It also facilitates the use of linguistic reasoning. It is based on the premise that traditional sharp (crisp) boundaries (on/off, yes/no etc) are inadequate in dealing with real life systems. Gradual memberships of sets are therefore introduced. This is intuitively easy to accept since concepts such as 'quite full' or 'fairly long' are part of everyday speech. Thus the River Swale might fall somewhat into the set of long rivers but less so than the Amazon. It differs from probability in that fuzziness (or degree of membership) remains, even as events are realised. Thus one might estimate the probability of rain tomorrow at 50 per cent (or 0.5) which then is realised or not. However, a fuzzy estimate of membership of rainy weather of 50 per cent would constitute a somewhat rainy situation ie not heavy rain, probably mostly cloud with a little precipitation.

The concept of degree of membership can be illustrated by reference to tall people. There is no single height above which people may be classified as tall. It is therefore a fuzzy concept. An arbitrary figure, say 1.75metres could be adopted, whereby everyone of greater height is tall and everyone below is non-tall. However, there is clearly a continuum of heights and the boundary could be set wherever the assessor considers appropriate. Several linguistic terms could be used: tall, quite tall, not tall and each assigned a membership function, such as the distributions shown below.

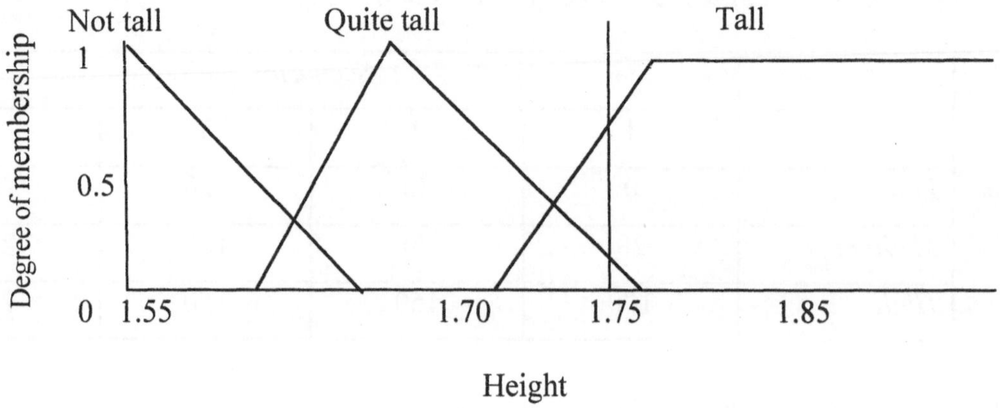

Figure 6.3 *Fuzzy Sets – Tall People*

Someone who is 1.75 metres might therefore be considered as belonging 0.6 to the set of tall people and 0.2 to the set of quite tall people.

Environmental examples of essentially fuzzy concepts could be the historic use of trigger and action values in dealing with contaminated land with mid-values where 'professional judgement' was deemed to be appropriate. An example of 'forcing' a crisp value would be the Seveso Directive, whereby a criterion for a notifiable accident causing immediate damage to the environment is considered to be one causing significant damage to 1 hectare or more of aquifer (significant however is a fuzzy concept). The implication of this might be that an incident involving damage to 0.99hectare is not reportable, however if an additional 100m^2 were affected, then it becomes significant.

In terms of practical application fuzzy sets are capable of mathematical manipulation (and being defuzzified if necessary to produce a crisp value) which facilitates the use of linguistic terms and of simplifying complex systems. Of course there is a subjective element in this in deciding the membership functions.

The discipline of management accounting has provided examples of the potential application of fuzzy logic to environmental issues (Lawrence et al, 1995). The first example discussed the many possible measures of toxicity reconciled to give a single function or value. This is particularly relevant to calculations such as those used in the RAGS assessment whereby two numerical values (for cancer and non-cancer risk) are derived from a whole range of diverse toxicity data, often extrapolated from high-dose animal work. Variables considered in the example include chronic/acute illness causing potential in humans, mutagenic potential in humans, potential to cause damage in non-human life forms and property damage causing potential. Assigned values such as high, moderate or very low could then be aggregated.

In the second example a fuzzy inference process is described to compare the potential damage associated with different wastes, where the environmental impact is considered to depend on toxicity and volume (but could be extended to include more variables).

A set of rules such as those below covered every combination of variables:

- if volume is high and toxicity is high then environmental impact is high.
- if volume is low and toxicity is high then environmental impact is modest.

Results from these assignments can be combined to give an overall set that is then defuzzified to give a single crisp value if required.

The type of approach described above might be usefully applied with formal management systems, where judgement is required on which environmental impacts are significant. This process is described in section 4.2.2 of ISO 14004, where impacts are first identified and then their significance is evaluated.

It would be possible to draw-up a questionnaire, which could be completed by an individual or a group, that would express (in linguistic terms) their judgement on issues relating to the overall assessment of significance.

Factors that would influence the significance test would include the following:

- scale of the impact (ie local, regional, global);
- size of the impact (eg large or small producer of hazardous waste);
- controls placed on the impact (eg management systems);
- uncertainty over the impact; and
- site sensitivity (proximity to local residents, sensitive ecosystems etc).

Several of the factors would need to be viewed together to assess significance, simple examples would be of the type:

If reasonably large quantities of oil/fuel are stored on the site *and* there is a drinking water well close to site *then* the risk of leaks is significant.

If very noisy processes are undertaken on the site *and* neighbours are located reasonably close *then* noise nuisance might be significant.

CHAPTER 7
RISK ASSESSMENT AND MANAGEMENT FRAMEWORKS

This chapter considers a number of risk-based frameworks intended for various environmental management applications. Many are presented with the idea that they may be useful in a situation beyond their original application or to demonstrate a particular aspect of environmental risk (eg sensitivity analysis). It cannot be overemphasised that they are all models of reality, not reality itself and that a simple, crude, approach might be more appropriate then a more complex approach in many cases.

Several models of varying complexity are described; in most cases the ranking or risk estimate is used as a basis for selective investment of money or management effort. As funding for remedial measures might not be available for all sites or options immediately it is advantageous to identify higher risks for priority action. All the systems described rely on expert judgement to different degrees, either in the formulation of the original model or in its application. In an extreme example an expert panel could be employed to rank sites. However, in most cases this would be impractical and a simple model would be adequate. Indeed the primary aim of the models is to encapsulate expert knowledge and predictions in a form that can readily be applied to real-life situations. An example would be the US Hazard Ranking System (HRS) in which a ranking model is applied to sites to determine future eligibility for central funding. The Environment Agency has utilised a similar approach (emphasising its subjectivity) to a classification of groundwater pollution from point sources in England and Wales, where they are ranked in severity codes from 1 (high significance) through to 5 (low significance) based on factors including the nature and extent of the contamination, the local hydrogeology (including aquifers) and remediation work undertaken.

The UK Royal Commission on Environmental Pollution (ENDS, 1998b) noted that even if a consensus exists among expert practitioners, other people using different (but equally valid) assumptions can give rise to very different risk estimates. It further noted that even where probabilities may be accurately estimated they may not, in themselves, be appropriate to arrive at a regulatory judgement (which incidentally might make one question the supposed inherent superiority of quantitiative over qualitative estimates). The question of subjectivity inherent in risk assessment is also considered in some detail by Waring and Glendon (1998). They note that although there is objectivity in that validated methodologies are used in risk-based models, the subjectivity arises through the choice of methodology in addition to the choice of elements such as assigned probabilities and projected outcomes.

The distinction that is often made between 'qualitative' and 'quantitative' risk assessment leads to the implicit assumption that the latter is inherently superior. However, the recognition of subjective aspects within all the models, means that it is difficult to draw a clear distinction, and none is offered in this publication.

In considering environmental risk assessment it is obvious that there are at least two distinct traditions from which the techniques have been developed. The first is the process engineering based approach, considering environmental risks as an extension of major accidents, the second, coming from a chemical/toxicological perspective, has developed from consideration of long-term exposure risks. Given the interdisciplinary nature of the subject and the importance of risk communication it will be instructive to see how the practitioners from different backgrounds embrace these wider perspectives in the future.

The use of risk assessment to deal with chronic risks such as those associated with contaminated land is complementary to the more traditional approach of considering the environmental consequences associated with catastrophic events. Two examples of these differing approaches would be the assessment required by the Seveso Directive and that conducted under the US Superfund regime (RAGS - Risk Assessment Guidance for Superfund). The implementation of the Seveso II Directive, Control of Major Accident Hazard Directive (COMAH) Council Directive 96/82/EC of 9 December 1996 on the control of major accident hazards involving dangerous substances (L10/13), represented an extension of the CIMAH regulations dealing with major safety hazards.

The UK has been particularly active in the study of safety risk assessment of major hazards as addressed by CIMAH/COMAH, whereby a numerical assessment of risk is based on consideration of exposure to a harmful agent and the dose-response behaviour of the human target. This approach is discussed here in purely environmental concerns. Clearly there is an area of overlap since human health and safety form a part of broader environmental concerns. The considerable body of available literature in this area such as the Health & Safety Executive (HSE) publication, *Risk Assessment of Major Hazards: A Brief Overview of Methods and Information Sources* (1989) should be consulted for more detailed discussion.

In many cases the purely environmental elements within CIMAH received cursory attention and the new environmental emphasis in COMAH should certainly act to promote improved environmental management. The definition of what constitutes a major accident to the environment under the regime relates to the notification criteria, more specifically, where a release leads to serious danger on or off site and involves specified dangerous substances. Categories of substances classified as 'dangerous for the environment' (R5O, R51, R53) now fall under the Regulations. In some ways this reflects a fuller consideration of the risks to the water environment, beyond the predominantly atmospheric-related issues considered earlier. Annex VI of the Directive states that the notification criteria for immediate damage to the environment included:

- permanent or long term damage to terrestrial habitats (0.5hectares of important habitat, 10 hectares of more widespread habitat);
- significant or long term damage to freshwater and marine habitats (eg 10km or more of river or canal); and
- significant damage to an aquifer (1 hectare or more).

Detailed consideration will need to be made of the recovery times of natural systems affected by accidents. Recent guidance (DETR, 1999a) has indicated that, for example, surface waters would normally anticipate re-establishment of prior quality quickly. Should contamination therefore persist for more than a few weeks it will certainly constitute a major accident to the environment.

Off-site emergency plans need to provide for the restoration and clean up of the environment following a major incident. Implicit in this is the need for some type of baseline review, possibly including a habitat survey and certainly identifying neighbouring features such as sites of Special Scientific Interest (SSSI).

Interestingly in the context of the communication issues discussed earlier, it has been suggested that COMAH compliance will become 'a regular if not permanent' feature of financial risk statements in annual reports (Waring, 1998b).

A COMAH type risk assessment is concerned with major accidents (ie high consequence/low probability events) and not routine legally sanctioned emissions, so it is therefore less relevant when dealing with environmental problems associated with less serious events that might occur over longer periods. Looking at chronic risks can be seen as broadening the spectrum from COMAH to consider high probability/low consequence situations that could be associated with contaminated land, for example.

The risk-based approach to contaminated land could involve initial ranking of sites (largely a desk exercise) which then identifies priorities for the detailed investigation for soil and groundwater contamination. A site quantified risk assessment can then help in developing defensible remediation standards. This approach has recently been acknowledged by the UK Government in terms of its strategy for dealing with contaminated land, reflected in the new provisions of the Environment Act, 1995.

A US Superfund RAGS risk assessment would consider the incremental cancer and non-cancer human health risks associated with (potentially) long-term exposure to various agents. A number of pathways might be considered, eg ingestion of contaminated drinking water, eating contaminated food or direct contact with contaminated soil. Having calculated the exposure, the risk can be calculated using factors in a large online database (IRIS).

Ideally, one would wish to account for harm to all components of an ecosystem and the Netherlands has attempted to accommodate this in their preferred methodology. However, uncertainties associated with this technique are likely to be even greater than when risk to humans alone is considered.

The models discussed here show a significant variation in their levels of complexity. There is a range running from relatively simple rankings of fuel storage locations through to the time and resource intensive methodologies used in Superfund. Nevertheless, it is clear that the underlying risk frameworks all show similarities and that there are significant assumptions made in all approaches. Some discussion is included that considers how the models might be adapted to apply to new situations divorced from their original (often regulatory) application.

Landfill Prioritisation

Environment Agency sponsored research on groundwater contamination by leachate from landfill sites has used a scoring system to identify priority sites (Environment Agency, 1995). Three categories were scored in an impact-ranking procedure based on the attributes of the aquifer (E), properties of the waste disposal facility (F) and the ground/surface water quality (W).

The site score (S) was calculated as the product of the environmental setting score (E) the facility score (F) and the water quality score (W).

$S = E \times F \times W$.

Scores for E, F and W were derived from:

$E = Ac + Uz + Fm + Ar$

Where: Ac = Aquifer class score
Uz = Unsaturated zone thickness score
Fm = Groundwater flow mechanism score
Ar = Annual average rainfall score

$F = Sd + Os + Sa + V + Wc$

Where: Sd = Site design score
Os = Operational status score
Sa = Site area score
V = Total waste volume score
Wc = Waste category score

$W = Gq + Pe + Pc + Sc + Pwg + Pws$

Where: Gq = Groundwater quality near site score
Pe = Plume extent score
Pc = Period of contamination score
Sc = Surface water contamination score
Pwg = Public Water Supply (PWS) groundwater source contamination score
Pws = PWS surface water source contamination score

Of particular interest was the sensitivity analysis undertaken. Different weightings were applied to each of the variables to assess whether one dominates the overall ranking.

$S = E^2 \times F \times W$
$S = E^{0.5} \times F \times W$
$S = E \times F^2 \times W$ etc

The scores were also obtained by summation:

ie $S = E + F + W$

The fact that there was consistency in the high scoring sites however the variables were combined, lent confidence to the idea that all the variables utilised in the model were significant.

Sensitivity exercises of this type could usefully be applied to many models, as in the experience of the author a number of models reflect principally a single variable. For example, a landfill model might see the overall risk correlating very strongly with the size of the site. Other variables intuitively considered to be important would therefore have little impact on the final results.

'Hazard' scores were also calculated as the product of the setting and the facility score (H = E x F). The hazard score is probably the most useful predictive measure of risk as groundwater monitoring is not uniformly adequate. A low quality score may therefore have reflected a lack of monitoring information. All the scoring systems identified 15 sites out of 102 studied that scored consistently highly. Problems with groundwater quality were anticipated at a further 22, where the inadequacies of monitoring meant that contamination was not observed.

Operator and Pollution Risk Appraisal

The revised version of the Environment Agency methodology – Operator and Pollution Risk Appraisal (OPRA) was issued in August 1997 following a broad consultation exercise. One of its principal aims was to target regulatory effort towards higher-risk operations thus improving Agency efficiency. It has been designed to complement authorisation and support inspections under IPC (and presumably under IPPC) as a simple screening tool. Although it has been designed for IPC processes it is capable of being applied more widely.

The evaluation process involves assigning scores to two groups of factors. Firstly, the factors inherent to a process such as material and scale of operations are considered. This is combined with location specific factors to give the 'pollution hazard'– effectively the inherent risk. Factors relating to the management of the process, such as maintenance and training, define the level of 'operator performance' in managing the pollution hazard. The overall risk is therefore determined from the combination of the pollution hazard and operator performance results.

Guidance notes are provided on scoring (on a range of 1–5) of the seven attributes of the Operator Performance Appraisal (OPA) and the Pollution Hazard Appraisal (PHA) as shown below.

Operator Performance Appraisal (OPA) Attributes:

1 Recording and use of information (2)
2 Knowledge and implementation of authorisation requirements (2)
3 Plant maintenance (5)
4 Management and training (4)
5 Process operations (5)
6 Incidents, complaints and non-compliance events (3)
7 Recognised environmental management systems

Pollution Hazard Appraisal (PHA) Attributes:

1 Presence of hazardous substances
2 Scale of hazardous substances
3 Frequency and nature of hazardous operations
4 Technologies for hazard prevention and minimisation
5 Technologies for hazard abatement
6 Location of process
7 Offensive characteristics

The bracketed numbers with the OPA attributes are weightings to be applied in recognition of the relative importance of each item. In evaluating OPA scores the following guidelines are given, a score of 1 equating to 'below requirements' further action may be needed, 3 is 'fully meets requirements' and 5 is 'above requirements in terms of actual environmental performance'.

As higher scores for both OPA and PHA means better management performance and higher hazard levels, a PHA/OPA plot would not show the same characteristics as most frequency/consequence graphs, where results near the origin are the lowest risk, as indicated in Figure 7.1. This does not invalidate OPRA but merely emphasises that the scales should be closely scrutinised in such an assessment to ensure misunderstandings do not arise.

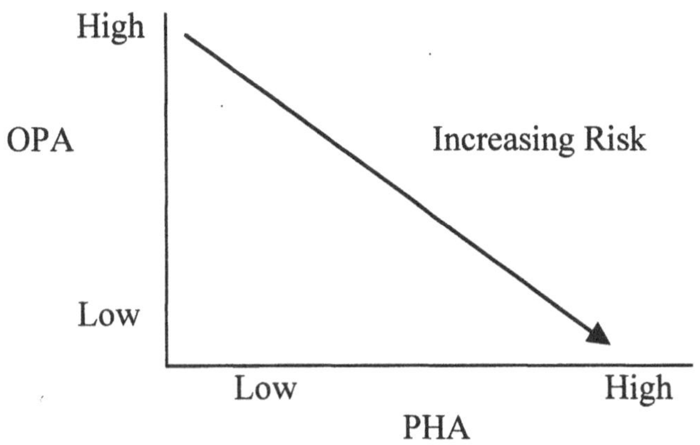

Figure 7.1 *OPRA - Increasing Risk*

In adapting OPRA to apply more broadly, some of the process authorisation specific elements would need to be changed. In addition, it could be worthwhile to review the weightings given to locations of process and offensive characteristics. The prominence given to offensive characteristics (odour etc) in OPRA probably reflects importance in terms of triggering complaints and regulatory involvement (including inspections). The site dependency of risks is considered in PHA6. Although some further guidance on site location is supported by elements in PHA1 and PHA5, it would hardly seem to justify their equivalence in terms of weightings.

The 'OPRA for Waste' draft methodology (DETR, 1999b) has recently been published and incorporates source and target aspects into its environmental appraisal (dependent on the type of facility, nature of waste, human dwellings and waste characteristics) this is complemented by an operator performance appraisal, made up of an assessment of non compliances and management control procedures already in place.

In conclusion, OPRA is a simple risk-based screening tool that has undergone considerable consultation and development. The Agency is currently expanding its scope to apply to other areas such as waste regulation. It can also be applied, notwithstanding some idiosyncrasies, by third parties relatively easily to almost any industrial operation

US Hazard Ranking System

The Hazard Ranking System (HRS) is a scoring system used by the US Environmental Protection Agency (EPA) and is the means by which the relative risks associated with the potential release of hazardous substances from waste sites are measured. The HRS score is the primary criterion used to determine whether a site is placed on the National Priorities List.

HRS is a model based on pathways. The score for each pathway is obtained by considering a set of factors that characterise the potential of the facility to cause harm. These factors are grouped into environmental categories that are then combined to give a score for each pathway. The pathway scores are then combined (using root mean squares) to arrive at the final ranking score.

In response to the Superfund Amendments and Reauthorisation Act (SARA), the HRS was reviewed and revised. The revision, amongst others, allowed for consideration of the potential for contaminated groundwater to enter surface water. The following discussion also considers an adaptation of HRS for use outside the US.

Four pathways can be scored in HRS:

1 groundwater migration (drinking water).
2 surface water migration (drinking water, human food chain, sensitive environments).
3 soil exposure (resident population, nearby population, sensitive environments).
4 air migration (population, sensitive environments).

The adapted risk methodology was developed on the basis of three migration pathways; 1 groundwater; 2 surface water; and 3 direct exposure. The scoring structure of each pathway was similar, and was expressed by the equation:

$$\left[\text{Pathway score}\right] = \left[\text{Likelihood of contamination release}\right] \times \left[\text{Material characteristics}\right] \times \left[\text{Exposed targets}\right]$$

The groundwater pathway evaluated the likelihood that hazardous materials at a site would migrate through the ground below the site and pollute aquifers and any groundwater abstractions that drew on those aquifers. The surface water pathway evaluated the likelihood that run-off containing hazardous materials from a site could move through surface water and affect people or sensitive environments. The direct exposure pathway evaluated the potential threats posed by direct exposure to hazardous materials or contaminated soil.

Each migration pathway consisted of the three categories (eg likelihood of release, material characteristics and exposed targets) represented by the aforementioned pathway score equation. These categories were further subdivided on the pathway score sheets and comprised five environmental categories: 1 observed release; 2 route characteristics; 3 containment; 4 material characteristics; and 5 targets.

The factors within the environmental categories were selected in order to rank the sites in terms of the potential risk they pose by describing:

- the manner in which the hazardous materials are contained;
- the route by which they would be released;

- the characteristics and amount of harmful materials; and
- the likely targets.

These factors are based on the operational and environmental data available for each site.

1 *Observed Release*: This category is scored if the measured concentration of the hazardous material is significantly above the background level and if that concentration could reasonably be attributed to the site. In the case of the groundwater and surface water pathways it relates to whether the site is known to have contaminated groundwater and surface water. In the case of the direct exposure pathway this category is scored if there has been a more detailed site investigation following an observed release.

2 *Route Characteristics*: This category addresses the route by which the hazardous material is able to reach potential targets. The factors considered include site geology and hydrogeology, soil type, proximity to water bodies, net effective rainfall and the physical state of the hazardous material.

3 *Containment*: This category is a measure of the engineering controls that are in place at the sites to prevent or minimise the release of hazardous materials. The factors which have been considered are the nature of the storage, whether it is above or below ground, the nature of the pipework, whether it is nylon or steel, and the presence of cathodic protection.

4 *Targets*: This category considers the targets potentially at risk from a release of hazardous material. Emphasis is placed on human targets and sensitive environments. Factors within this category relate to aquifer sensitivity, local groundwater and surface water abstractions, the presence of sensitive areas and sites of special scientific interest (SSSI).

In a formal HRS application the material characteristics (toxicity etc) would need to be considered, however, in the adapted model only a single material was considered.

Each factor within a pathway was assigned a numerical value or NS for no score if the information was not considered to be significant. This value was then multiplied by a weighting factor (the multiplier) yielding the factor score. The factor scores were then combined in the following way: factor scores within each environmental category are added, then the total scores for each of the five environmental categories were multiplied together to develop a score (multiplied total) for each of the three pathways. The pathway score was then normalised by dividing by the maximum possible pathway score and multiplying by 100. This puts the pathway score on a scale of 0 to 100. The normalised pathway scores were then combined as follows:

$$S = \frac{1}{1.73} \sqrt{Sgw^2 + Ssw^2 + Sde^2}$$

Sgw = groundwater pathway score.
Ssw = surface water pathway score.
Sde = direct exposure pathway score.

The results of combining the pathway scores in this manner accounted for the primary (highest scoring) pathway, while giving consideration to the secondary or tertiary pathways if they score high. The factor 1/1.73 simply reduces the final ranking score to a 100 point scale. Sites are then ranked on the basis of the total score, groundwater pathway score (Sgw), surface water pathway score (Ssw) or direct exposure pathway score (Sde) according to requirement. HRS is probably the most highly developed of any risk-ranking model and has had the benefit of considerable practical experience in its application.

It has certainly not been immune to criticism, Taylor (1995) for example, described problems with HRS being 'legion'. He considered the fact that the single most hazardous substance is used to score the toxicity for the site and that greater weight is given to the number of people near a site than to the toxicity of the hazardous substances, to be the principal problems. In making any adaptations to the model it must be recognised that changes could invalidate as well as improve the model. Software (PREscore) is now available from the USEPA which aims to speed up the scoring process and reduce the number of errors arising.

Underground Storage Tank Rating

Leaking underground storage tanks can result in significant contamination as a leak can go undetected for many years and this has lead to strict testing requirements in the US. Siting underground makes direct inspection and maintenance of the tanks more difficult. The US risk model described here (Casana and Dixon, 1989) involves a rating system based on frequency of release (leak) and consequence. On the basis of the two rating factors the tank can be assigned to different risk categories varying from low risk, through moderately low, moderately high to high risk (Table 7.1). The 'release potential' equates broadly to the 'frequency' criterion in other models.

Table 7.1 *UST Release Potential Factors*

	Low Risk	*Moderate Risk*	*High Risk*
Tank System Age –Yrs	0 – 10	11 – 20	21+
Corrosion Protection	Tank and/or piping protected	Either tank or piping protected	None
Leak Detection	Continuous	Intermittent	None
Soil Corrosivity	pH>6.5	pH 5.5. – 6.5	pH<5.5

Table 7.2 *UST Damage Potential Factors*

	Low Risk	*Moderate Risk*	*High Risk*
Depth to water table (ft)	> 50	15 – 20	< 15
Soil hydraulic Conductivity (gal/day/ft^2)	< 1 (eg clay)	1 – 10^3 (eg sand)	>10^3 (eg gravel)
Local population served by public groundwater supply	< 1,000	1,000 – 10,000	> 10,000
Distance to surface water (ft)	> 1,000	< 1,000 (other than public water supply)	< 1,000 (public surface water supply)

In terms of allocating points for each factor, no points are accrued for each low risk item, one point for each moderate risk and two for each high risk factors. The damage potential factors were scored as described in Table 7.2.

The range of available scores applied (to each tank individually) is zero to ten for both damage and release potential. The values are then placed on a risk-rating chart (Figure 7.2).

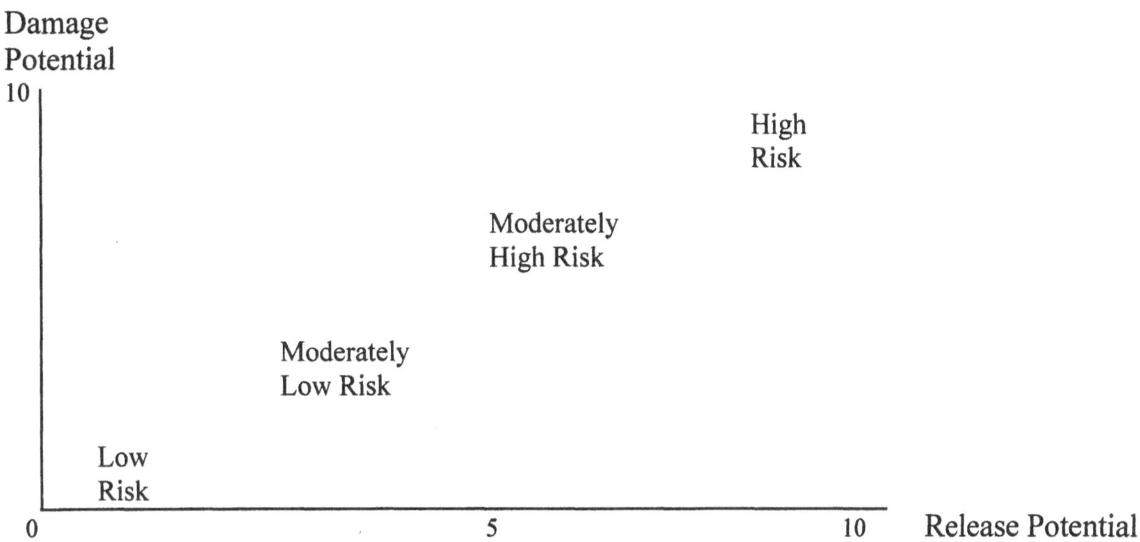

Figure 7.2 *Risk Rating Chart*

The tanks in the bottom left hand corner represent the lowest risk tanks: they have a low potential to release and their surroundings are of a low environmental sensitivity. Moving towards the top right hand corner results in increasing bands of risk.

Fuel Storage Risk Ranking

The application of a simple model capable of being used by non-specialists to rank fuel storage facilities is described here. It could, nonetheless, provide sufficient information for the purpose of deciding where to focus funds. The example relates to an organisation with a large number of above ground storage tanks at a variety of locations around the UK. A routine audit at several sites indicated that the tanks had containment capabilities of variable quality and might pose significant environmental risks. The organisation wished to upgrade the containment at all sites; however, it could only instigate improvement works at approximately 20 per cent of the sites each year. A risk ranking enabled those sites considered to pose the highest risk to receive priority attention.

The risk analysis was based on probability release and consequence factors that were combined to produce an overall risk factor for a particular site. The two basic premises underlying the risk ranking model are as follows:

1 It is assumed that the greatest environmental risk posed by operational sites is from the accidental release of fuel to ground or surface water. Environmental sensitivity of sites is therefore related to proximity to surface watercourses and major aquifers.

2 The probability of a release of fuel from a site is related to the quantity of fuel held and the current condition of containment measures. In addition, management and operational controls are considered as important elements, and therefore the risk factor allocated to a site could be reduced by improving procedures and the awareness of site staff.

A simple scoring system allocates sites a score of 1 – 4 (for probability release factors) and 1 – 5 (consequence factors). A below average risk is represented by a low score and an above average risk is represented by a high score. The individual scores for probability release factors and consequence factors are summed separately, and the multiplication of the two provides a total risk factor score for each site. An example scoring for two sites is shown in Table 7.3.

Table 7.3 *Environmental Risk Ranking – Site Risk Factors*

	Site A	Site B
Probability Release Factors:		
Quantity of Fuel	4	3
Bunding	3	3
Drainage	3	3
Interceptors	3	3
Contingency	2	3
Total	15	15
Consequence Factors		
Aquifer	2	4
Watercourse	3	4
Total	5	8
TOTAL RISK FACTOR	75	120

A panel of individuals (including two people who had conducted site audits) then considered a selected sample of sites, and based on all available information, decided on where the boundaries should be set for classification as above average, average or below average risk. Cross reference was then made to the site risk scores and used to set the boundary scores for the three categories:

- total risk factor score of > 100 represents an above average risk;
- total risk factor score of 70 – 100 represents an average risk; and
- total risk factor score of <70 represents a below average risk.

Probability Release Factor

A Quantity of Fuel on Site:

1 No fuel containing equipment on site
2 Less than 250,000 litres of fuel
3 Between 250,000 and 500,000 litres of fuel
4 More than 500,000 litres of fuel

B Condition of bunds/containment:

1 Bunds in good visual condition and appropriate for the purpose
2 Equipment bunded but poor seals and possible gradual leaks
3 Equipment bunded but bunds probably not retaining water, and/or bunds single skinned, and/or bunds considered too small
4 No containment or bunding of equipment

C Drainage:

1 Up-to-date drainage plans available, drains and outfalls included on inspection routines and system well maintained
2 Drainage plans available and staff aware of layout and outfalls but system not maintained
3 No drainage plans available, staff unaware of layout and drains blocked
4 Visible fuel in drains or at outfall

D Interceptors:

1 Interceptors visually well maintained
2 Interceptors adequately sized but poor maintenance
3 Interceptor inadequately sized and/or storm bypass
4 No interceptor

E Contingency:

1 Contingency plans in place, including spill response equipment and staff awareness of procedures
2 Contingency plans in place and staff awareness but no spill response equipment
3 Contingency plan but no staff awareness or spill response equipment
4 No contingency plans

Consequence Factors

A Groundwater Vulnerability:

1 Non-aquifer
2 Minor aquifer not used for public water supply
3 Major aquifer – soils of low permeability and therefore unlikely to permit passage of pollutants
4 Major aquifer – soils of moderate permeability
5 Major aquifer - soils of high permeability

B Watercourses:

1 No drains on site
2 Drainage to sewer
3 Drainage discharges to surface waters of poor quality

4 Drainage discharges to surface waters of fair quality
5 Drainage discharges to surface waters of good quality or the site is located within 50 metres (approx) of a watercourse of similar quality

Risk-based Corrective Action

The demand for a flexible framework to deal with contaminated sites in the US led to the development of the 1998 ASTM Standard PS104-98 Standard Provisional Guide for Risk-based Corrective Action (RBCA – usually pronounced Rebecca). It is intended to provide a technically defensible framework allowing a progression from simple techniques to more complex, quantitative risk assessments as required. Its ambitious aim is to complement regulatory requirements, integrating both human health and ecological concerns in a system that encourages stakeholder involvement in the decision making process. Unlike some other US standards, the guidance is sufficiently generic to envisage that it can be usefully applied outside the US with little modification. A tiered 1-2-3 approach is adopted, with the need to clarify technical policy decisions at the start of the process (ie recognising the site specific regulatory, stakeholder, technical and other requirements including planned future use at the outset).

Sites are initially classified according to the urgency of response required. This initial site assessment draws on available site information including chemicals of concern, targets and pathways. The site is then classified to ensure that sites posing an immediate threat receive appropriate prompt attention. Example site classifications given are:

1 Immediate threat to human health, safety or sensitive environmental receptors
2 Short term (0–2 years) threat
3 Long-term (greater than 2 years) threat
4 No demonstrable long-term threat

The first part of the tier 1 process involves the development of a conceptual site model in which potential sources, transport pathways and receptors are identified. Should there be relevant complete exposure pathways, comparisons are then made of chemical(s) of concern with generic criteria. Human health concerns are addressed in tier 1 by comparison with non-site specific Risk-based Screening Levels (RBSL). The look-up RBSL tables for soil, groundwater, sediment, surface water and vapour will be typically derived from USEPA toxicological data (as per RAGS), however could also accommodate particular local or regulatory concerns.

Source area concentrations are compared directly with the RBSL in tier 1 (ie no account is taken of transport effects) for identified exposure pathways. The standard does not anticipate that quantitative ecological criteria will be utilised in all cases, however, generic Relevant Ecological Screening Criteria (RESC) in the tier 1 evaluation may, nonetheless, include biological measures or chemical concentrations. An ecological assessment is therefore likely to comprise an initial identification of relevant receptors and habitats on or near the site. Should potential exposure pathways exist, site specific assessments will be required. Should site conditions fall below the screening criteria no further action might be justified, conversely, RBSL might be used as remediation targets in the case of elevated concentrations. A tier 2 evaluation may be triggered where Site Specific Target Levels (SSTL) and Site Specific Ecological Criteria (SSEC) are determined. This is achieved by replacing the non-site specific assumptions (including exposure aspects) used in the tier 1 analysis. A decision process similar to that employed in the tier 1 assessment indicates what subsequent actions are required. A tier 3 analysis might be started which is likely to require significantly greater data requirements and more complex analysis (eg using sophisticated fate and transport models). The fundamental aim of all the analyses is to arrive at a suitable remedial action programme.

Current and reasonably anticipated future uses of the site and its surroundings should form part of the evaluation. Should site conditions change eg from commercial to domestic use, SSTL values will more than likely need to be reviewed and recalculated as appropriate.

Should tier 1 look up tables of RBSL not be available, the user is expected to develop something suitable. An appendix to the standard contains an example of developing table values for benzene, chlorobenzene, mercury and other chemicals of concern. For example, the equation below has been used to develop the RBSL for soil relating to leaching to groundwater:

$$\text{RBSL} = \frac{\text{RBSLgw}}{\text{Lfsw}}$$

RBSL = Risk-based Screening Level for soil (mg/kg)
Lfsw = Leaching factor, soil to groundwater (g-soil/cm^3-water)
RBSLgw = Risk-based Screening Level for groundwater ingestion (mg/l)

Where:

$$\text{Lfsw} = \frac{\text{Lfpw,gw}}{\text{Ksw}}$$

Lfpw,gw = Leaching factor, soil pore water/groundwater concentration ratio (cm^3-water/cm^3-water)
KSW = Soil to water partition coefficient (cm^3-water/g-soil)

The problems using commonly determined multicomponent measurements such as total petroleum hydrocarbons (TPH) and total organic carbons (TOC) are emphasised in the standard. Where a single value is used to represent a class of compounds it is considered that they provide insufficient information on individual compounds to facilitate a calculation of risk.

Guidelines on public involvement in the process including communication are also described. This might involve a drafting of a public involvement plan, including a background history of the site, clean-up already taking place and opportunities for public participation in clean-up decisions. The importance of risk communication with residents (as often as possible, and at a level that can be easily understood) is also emphasised. Finally, the importance of monitoring is underlined as a means to:

- demonstrate the effectiveness of remedial action;
- confirm that current conditions persist or improve with time; and
- verify model assumptions and conditions.

Dutch Contaminated Land Framework

In order that the risks associated with the exposure of ecosystems to chemical contaminants could be accommodated within a legal framework, the Dutch National Health Council evaluated various ecosystem risk assessment strategies. This led to a considerable amount of research in the Netherlands on quantitative methods for ecosystem risk assessment linked to contaminated land.

The Dutch contaminated land regime is founded on risk-based criteria utilising both human health and ecosystems risk elements (Swartjes, 1997). Two sets of values termed Target and Intervention are calculated, allowing classification of soils as clean, (slightly) polluted and seriously polluted. The Target Values are based on potential ecosystem risks and the Intervention Values on both potential ecosystem and human health risks (Intermediate Values might also be calculated in some cases which are simply the mean of the Target and Intervention Value). The values are not dependent on site use, ie they are multifunctional.

The human exposure models are used in the calculation of standards based on exposure to terrestrial soils (CSOIL) and exposure to sediment (SEDISOIL) in exposure routes such as direct ingestion, drinking water intake and inhalation during bathing for soils. Fish consumption and surface water ingestion are considered for sediments. The effects assessment considers threshold (eg genotoxic carcinogens) and no-threshold contaminants differently. The Maximum Permissible Risk for Intake (MPR human) is defined as the dose which causes a risk (in the case of genotoxic carcinogens) of one additional case of lethal tumour in 10,000 lifelong exposed individuals.

In ecosystems risk assessment two aspects considered are irreparable damage to species composition and adverse effects on microbial and enzymic processes. Should not enough terrestrial data be available, aquatic phase data can be used, combined with suitable partition coefficients.

The method adopted for soil ecosystem risks involved determining the concentrations of a toxic substance below which there were no observable adverse effects (NOAEC) on a range of representative soil-living organisms. These values are then assumed to form part of a statistical distribution such that it is possible to derive a relationship between the percentage of species in an ecosystem that are experiencing an excess of their NOAEC values and the concentration of a toxic substance in the soil. Using the above relationship, the maximum permissible risk level for a particular substance was chosen as that which fully protects 95 per cent of the species in the ecosystem.

The Target Values are related to 'negligible risk' to ecosystems, assumed to be 1 per cent of the maximum permissible risk (defined as the concentration providing 95 per cent protection – HC95). The Ecological Intervention Value is based on the 50 per cent protection value (HC50). Soil quality resulting in exceedance of the Maximum Permissible Risk for intake (MPR human) is set at the Intervention Value (note this is the potential exposure using standard conditions). All exposure pathways within CSOIL are therefore included in the 'standard' calculation.

The lower (ie more stringent) of the ecological and human health intervention value is the one that is adopted. Intervention Values for groundwater are derived from Soil Intervention Values. Uncertainty scores are included for both ecotoxicological and human health values. Some examples of values are shown below in Table 7.4. The higher the Uncertainty Score (US) the lower the uncertainty in the quoted values.

Several problems become apparent when applying methods such as those described above for ecosystem risks, over and above the uncertainties that might be involved in estimating human health risks. The validity of the extrapolation from selected indicator species to an overall system might

Table 7.4 *Dutch Intervention Values*

	Ecotoxicological Int. Values	*US*	*Human MPR Value (mg/kg/day)*	*Human Int. Value (mg/kg)*	*US*	*Integrated Int. Value Soil (mg/kg)*
Arsenic	40	6	2.1	680	3	40
Lead	290	10	3.6	300	3	290
Benzene	25	2	4.3	1.1	6	1.1

be questioned; however, of potentially greater concern might be the lack of chronic ecotoxicological data and the poor understanding of population level effects. For example, the presence of a toxic agent might favour more resistant individuals within a species, which would come to dominate and the overall population is little affected even though the concentration was greater than the NOAEC.

Superfund Risk Assessment

The primary objective of the US Superfund programme is to protect human health and the environment from current and potential threats posed by uncontrolled releases of hazardous substances. The *Human Health Evaluation Manual* describes the process of gathering information and assessing the risk to human health and is part of Risk Assessment Guidance for Superfund (RAGS). The *Human Health Evaluation Manual* has several parts: Part A describes the baseline risk assessment, Parts B and C deal with refinement of preliminary remediation goals and risk evaluation of remedial alternatives. Part D has been implemented in 1998, providing templates for presentation of calculations etc to improve consistency and transparency. From a UK perspective the interesting elements are that the process shows a quantifiable method of describing and assessing chronic environmental risks associated with hazardous materials. It provides an interesting counterpoint to the COMAH approach focusing on major accidents. It is also helpful in understanding the process by which contaminant levels in soils could be set as the basis for remediation efforts. Although considerable effort would need to be made in terms of analysis to enable the derivation of site-specific clean-up standards, it would clearly be worthwhile on larger sites where significant clean-up costs might be incurred for uncertain benefit.

The two major criticisms levelled at RAGS are that 1 it is too conservative in its estimates; and
2 the strength of the toxicological data and its extrapolation to human effects.

Criticism 1 has been commonly voiced in the US, and centres on the fact that the method uses reasonable maxima rather than most probable estimates. The resulting final estimate is therefore likely to overestimate significantly the actual risk posed by contaminants.

Criticism 2 is much more familiar in the UK. Concern over the extrapolation of animal toxicity data applying to human risks has been voiced widely. The process itself incorporates uncertainty factors and the uncertainty involved could rise to orders of magnitude. The criticism probably arises out of the fact that assumptions are made explicit in the process, which of course is highly desirable but does act to emphasise the degree of uncertainty. It was difficult to constructively criticise values such as the trigger levels for contaminated land used historically in the UK when the underlying assumptions were not transparent. The ecotoxicological derived values utilised in the Netherlands also utilise assumptions when extrapolating up from individuals of selected indicator species to effects on overall soil systems. This poses particular problems because in ecosystems (as opposed to humans) the emphasis is much more focused on protection of the overall population rather than the individual. Part A of the *Human Health Evaluation Manual* deals in a prescriptive manner with the baseline assessment of risks, handling carcinogenic and non-carcinogenic risk separately.

Superfund Risk Assessment: Part A

EPA/540/1-89/002 Risk Assessment Guidance for Superfund Volume 1: *Human Health Evaluation Manua*l (Part A) pulls together all the requirements for a Superfund Risk Assessment including risk characterisation. This manual, developed for use at hazardous waste sites, presents a process of gathering and assessing human health risk information. Guidance is given on appropriate methods and data. The baseline risk assessment process described in Part A of the manual consists of the following main steps. Relevant information identified through data collection and evaluation is used to develop exposure and toxicity assessments. Risk characterisation summarises and integrates both the toxicity and exposure steps into quantitative and qualitative expressions of risk. In each of the steps in the Part A risk assessment process, data is gathered and interpreted and inferences are made about risk factors. Often the interpretation of the information is expressed as the best scientific judgement on the part of the risk assessor.

Data Collection

This describes procedures for acquiring reliable chemical release and exposure data for quantitative assessment. The topics discussed in the Data Collection chapter are shown below:

- available site information;
- modelling parameter needs;
- background sampling needs;
- preliminary identification of human exposure; and
- overall strategy for sample collection.

Data Evaluation

This describes the steps involved in organising data and identifying a set of chemicals and concentrations that are of acceptable quality for use in the quantitative risk assessment. The data evaluation steps are shown below.

Data Evaluation Steps
- Gather all data available from the site investigation and sort by medium (water, soil etc).
- Evaluate the analytical methods used (are they acceptable in terms of specificity, accuracy and sensitivity?).
- Evaluate whether tentatively identified compounds should be included in the risk assessment.
- Compare potential site-related contamination with background levels.
- Finalise the data set for use in the risk assessment, limiting, where appropriate, the number of chemicals to be carried through the risk assessment. Where compounds contribute little to the overall toxicity (eg < 1 per cent) they can be effectively screened out thus simplifying the data handling and assessment (and further sampling analysis).

Exposure Assessment

This step involves consideration of the site setting and pathways in order to calculate exposure at the chosen receptors. Standardised assumptions on parameters such as inhalation rate and body weight are combined with site-specific details. The EPA approach is to calculate Reasonable Maximum Exposures (RME). This approach is considered to address the inherent uncertainty in the whole process. For example, rather than use the average concentration of chemical contacted over the exposure period, the 95 per cent upper confidence limit value is used. Conservative values are also used for exposure frequency and duration. Computer-based modelling of the movement of contaminants in air and groundwater is often undertaken at this stage in order to predict future exposure patterns.

Toxicity Assessment

Toxicity information for the target compounds is collected using the hierarchy of sources identified in the manual. The Integrated Risk Information System (IRIS) is an online database that is viewed as the primary source of information. It contains cancer and non-cancer human health risk information on over 500 chemicals. The EPA recommend that reference is made to the periodically updated Health Effects Assessment Summary Tables (HEAST) for chemicals not addressed in IRIS.

Risk Characterisation

This stage represents the culmination of the risk assessment in which the results from the exposure and the toxicity assessment are brought together. The risk from individual chemicals is assessed in terms of a hazard index for non-carcinogens and incremental cancer risk for carcinogens. These values are then combined separately to quantify the carcinogenic and noncarcinogenic risks arising from exposure to multiple chemicals.

An example of calculation for the carcinogenic risk associated with a single chemical via vapour phase inhalation is as follows:

Example Calculation Inhalation of Vapour Phase Chemicals

$$\textit{Intake (mg / kg day)} = \frac{\text{CA x IR x ET x EF x ED}}{\text{BW x AT}}$$

CA = Analyte Concentration (mg/m^3)
IR = Inhalation Rate (adult) (m^3/h)
ET = Exposure Time (h/day)
EF = Exposure Frequency (days/yr)
ED = Exposure Duration (years)
BW = Body Weight (kg)
AT = Average Time

Risk is then characterised using the following relationship shown below where the slope factor has been obtained from IRIS or similar sources.

Risk (unitless) = Daily Intake (mg/kg day) x Carcinogenic Slope Factor $(\text{mg/kg day})^{-1}$

Superfund Risk Assessment: Part A

Recently produced supplemental guidance to Part A is arrived at promoting the involvement of local communities in the Superfund decision making process. It recognises that community input can help the process in several ways:

- identifying overlooked local knowledge;
- ensuring that special issues or concerns are included at the outset; and
- improve acceptance of the process.

Superfund Risk Assessment: Part B

Part B of the *Human Health Evaluation Manual* describes how to develop risk-based preliminary remediation goals (PRG). These are clean-up targets for individual chemicals using standard default equations.

An example of an equation used relating the consumption of contaminated drinking water is as follows (based on an acceptable incremented cancer risk of 10^{-6}):

$$\text{Risk-based Preliminary Remediation Goal (PRG) (mg/1)} = \frac{1.7 \times 10^{-4}}{2(\text{SFo}) + 7.5\,(\text{SFi})}$$

where:

SFo = oral slope factor (mg/kg day)
SFi = inhalation slope factor (mg/kg day)

This equation accounts for two modes of exposure for drinking water, firstly, ingestion and secondly, inhalation of volatile contaminants from the water.

Human exposure assumptions for residential and industrial scenarios are considered. The exposure pathways and media considered for residential land use are:

- Ingestion and inhalation of contaminants in surface water or groundwater used for domestic supply; and
- Ingestion of contaminants in soil.

For industrial/commercial land use the exposure pathways are: ingestion of contaminated soil, inhalation of airborne particulates and volatile contaminants from soil.

Direct exposure to gamma radiation from soil contamination is considered by both land use scenarios.

The process essentially starts from the point of acceptable incremental risk. For carcinogenic risks an increment of 10^{-6} and for non-carcinogens a hazard index of unity are the starting points. The calculation then works in the opposite direction to the Part A assessment and B uses a pathway specific exposure assumption to calculate a maximum acceptable soil concentration.

Superfund Risk Assessment: Part C

Part C of the *Human Health Evaluation Manual* is concerned with evaluation of remediation alternatives. It deals with both long-term (ie residual) risks and short-term risks (ie those which arise through remedy implementation). The risk assessment may be either quantitative (paralleling the process in Part A) or qualitative.

A quantitative assessment would follow the same approach as described in Part A. The decision as to whether to conduct a qualitative or quantitative evaluation of remedial alternatives will depend on a number of factors including the 'perceived risk' associated with the alternative. The 'perceived risk' includes the professional judgement of scientists and engineers along with local community concerns. Factors that are said to lead to a generally higher level of perceived risk include:

- close proximity of populations;
- use of an on-site incinerator (for destruction of toxic material);
- large uncertainties in the nature of releases, as might occur with innovative technologies; and
- Multiple releases occurring simultaneously (eg from technologies operating in close proximity).

Superfund Risk Assessment: Part D

The aims behind Part D of RAGS are to improve the presentation of Superfund risk assessments. By provision of Standard Tools it is intended that the process will have greater transparency, clarity and consistency.

Standard tools presented in Part D include a Technical Approach for Risk Assessment (TARA) and Standard Tables with associated instructions. TARA is a ‘road map’ for continuous involvement of the regulators to facilitate decision making. Standard Tables are basically electronic templates to document important data, calculations and conclusions.

ACRONYMS AND ABBREVIATIONS

AIRMIC	Association of Insurance and Risk Managers
ASTM	American Society for Testing and Materials
BOD	Biochemical Oxygen Demand
BGS	British Geological Survey
CBI	Confederation of British Industry
CERCLA	Comprehensive Environmental Response, Compensation and Liability Act (US)
CFC	Chlorofluorocarbons
CIMAH	Control of Industrial Major Accident Hazards
CIRIA	Construction Industry Research and Information Association
COD	Chemical Oxygen Demand
CLEA	Contaminated Land Exposure Assessment
COMAH	Control of Major Accident Hazards
COSHH	Control of Substances Hazardous to Health
DETR	Department of the Environment, Transport and the Regions
EA	Environment Agency
EBRD	European Bank for Reconstruction and Development
EC	European Commission
EIL	Environmental Impairment Liability (Insurance)
EMAS	Eco-management and Audit Scheme
EU	European Union
HAZID	Hazard Identification Study
HAZOP	Hazard and Operability Study
HEAST	Health Effects Assessment Summary Tables (US)
HML	High, Medium, Low
HRS	Hazard Ranking System
HSE	Health and Safety Executive
ICRCL	Interdepartmental Committee on the Redevelopment of Contaminated Land
IPPC	Integrated Pollution Prevention and Control
IRIS	Integrated Risk Information System (US)
ISO	International Standards Organisation
LC50	Lethal Concentration (the concentration of a toxic substance that kills one half of a group of test animals in a given period).
LPC	Loss Prevention Council
LD50	Lethal Dose (the dose of a toxic substance that kills one half of a group of test animals in a given period).
NOAEC	No Observable Adverse Effect Concentration
OPA	Operator Performance Appraisal
OPRA	Operator and Pollution Risk Appraisal
PCB	Polychlorinated biphenyls
PHA	Pollution Hazard Appraisal

PI	Professional Indemnity
POST	Parliamentary Office of Science and Technology
PRG	Preliminary Remediation Goal (US)
RAGS	Risk Assessment Guidance for Superfund (US)
RBCA	Risk-based Corrective Action
RESC	Relevant Ecological Screening Criteria
RICS	Royal Institution of Chartered Surveyors
RME	Reasonable Maximum Exposure
SARA	Superfund Amendments and Reauthorisation Act (US)
SSEC	Site Specific Ecological Criteria
SSSI	Site of Special Scientific Interest
SSTL	Site Specific Target Levels
TOC	Total Organic Carbon
TPH	Total Petroleum Hydrocarbons
USEPA	United States Environmental Protection Agency
UST	Underground Storage Tank

REFERENCES

ABI (1998) Joint Pollution Working Group Recommendations for the Underwriting of Pollution Risks

Aon Risk Services (1999) Biennial Risk Financing and Insurance Survey

APS (1996) *Guidelines for Managing Risk in the Australian Public Service*, Joint Publication of the Management Advisory Board and its Management Improvement Advisory Committee Report No 22, Oct 1996

ASB (1998) *Financial Reporting Standard 12, Provisions and Contingencies*, Accounting Standards Board

ASTM Standard E1527-97 (1997) Standard Practice for Environmental Site Assessments: Phase I Environmental Site Assessment Process.

ASTM Standard E1528-96 (1996) Standard Practice for Environmental Site Assessments: Transaction Screen Process,

Barker, I (1997) Journey into the Next Century for Risk Managers, *AIRMIC Express*, October 1997, pp16–17

Bennink, D (1998) Superfund Solutions, *Risk and Insurance*, July 1998, pp40–44

Bernstein, P L (1996) *Against the Gods*, John Wiley & Sons Inc, New York

Borden, S and Sarkar, A (1996) Securitising Property Catastrophe Risk, *Current issues in Economics and Finance*, 2, 9, August 1996 pp2–6

Bridge, P (1997) Managing Environmental Risks on a Multi-site Basis, *AIRMIC Newsletter* March 1997, pp6–9

Business Risk (1999) *Convergence of Capital Markets.* January 1999, pp14–15

Cahill, L B (ed) (1989) *Environmental Audits* 6th Edition, Government Institute

Calow, P (1997) *Handbook of Environmental Risk Assessment and Management*, Blackwell Science

Casana, J and Dixon, G (1989) Application of quantitative risk assessment evaluation of underground storage tanks to insurance, banking and real estate transactions, in *Petroleum Contaminated Soils 2* (ed E J Calabrese), Lewis Publishers

CBI (1998) *Worth the Risk*

Cook, S (1992) Waste Hazards in *Environment Risk,* July/August pp37–38

Croner (1997) Environmental Insurance, *Croners Environmental Briefing* No 65

Davies, P (1999) A Fine Industry, *The Chemical Engineer*, 681, 27th May, pp22– 23

Dembo, R S and Freeman, A (1998) *Seeing Tomorrow: Rewriting the Rules of Risk*, John Wiley and Sons Inc, New York

DETR (1998) *UK Implementation of EC Directive 96/61 on Integrated Pollution Prevention and Control Second Consultation Paper*, January

DETR (1999a) *Guidance on the Interpretation of Major Accidents to the Environment for the Purposes of the COMAH Regulations*, Stationery Office, June

DETR, (1999b) Waste Management Licensing, Risk Assessment Inspection Frequencies, Operator Pollution Risk Appraisal *OPRA for Waste*, A Consultation Paper, April 1999

DoE (1991) *Interpretation of Major Accident to the Environment for the Purposes of the CIMAH Regulations, June 1991,* Department of the Environment

DoE (1995) *A Guide to Risk Assessment and Risk Management for Environmental Protection,* Department of the Environment

Economist Intelligence Unit (1996) *Managing Business Risks: An Integrated Approach*

Edwards, D (1998) *The Link Between Company Environmental and Financial Performance*, Earthscan Publications Ltd, London

Environment Agency (1995) *The Effects of Old Landfill Sites on Groundwater Quality: Phase I*, Environment Agency R&D Note 415

Environment Agency (1998) Water Pollution Incidents in England and Wales 1997

Environment Agency (1999) *Managing Firewater and Major Spillages,* Pollution Prevention Guidelines, p18

Environment Business (1997) *EMS System cuts fine by £70,000,* 6 November p6

Environment Business News Briefing (1999a) 28 Jan 1999, pp11–12

Environment Business News Briefing (1999b) 25 Feb 1999, p6

Environmental Analysis Co-operative, (1999) *Don't forget the Environment: A Guide for Incorporating Environmental Assessment into your Project*, IChemE

ENDS (1995) *Directory of Environmental Consultants*, 4th Edition

ENDS (1997a) *ENDS Report* 269, June 1997, p46

ENDS (1997b) *ENDS Report* 272, Sept 1997,p15

ENDS (1997c) *ENDS Report* 269, June 1997, p9

ENDS (1998a) *ENDS Report* 287, Dec 1998, p2

ENDS (1998b) *ENDS Report* 286, pp20–23

Epstein, M J (1996) *Measuring Corporate Performance*, Irwin Professional Publishing, Chicago

European Bank for Reconstruction and Development (1996) *Environmental Procedures (1992, revised 1996),* London

FT (1998), *Financial Times* 12/13 September 1998

Ferguson, C and Denner, J M (1994) Developing Guidelines (Trigger) Values for Contaminants in Soil: Underlying Risk Analysis and Risk Management Concepts in *Land Contamination and Reclamations* 2, 3, pp117–123

Ferguson, C and Denner, J M (1998) Human health risk assessment using UK Guideline Values for contaminants in soils – in Lerner, D N and Walton NRG (eds) Contaminated Land and Groundwater: *Future Directions*, Geological Society Special Publication 14, pp37–43

Figge, F (1998) *Systematisation of Economic Risks through Global Environmental Problems: A Threat to Financial Markets*. Sarasin Basic Report, July 1998

Fink, S (1998) A new measure of damages for environmental contamination, *The Safety and Health Practitioner,* October, p70

Finley, B and Paustenbach, D (1998) 'The Benefits of Probabilistic Exposure Assessment: Three Case Studies Involving Contaminated Air Water and Soil' in *Risk Analysis* 14, 1, pp53–73

Gowland, RT (1993) 'A risk management system of quality, safety and environmental affairs' in *Manosaf 93 1ChemE Symposium,* 10–11 February 1993, pp. 61–82.

Harris-Jones, J (1998) *The Management of Corporate Risk: A Framework for Directors*, The Association of Corporate Treasurers

HSE (1994) *Health and Safety Executive 5 steps to Risk Assessment* IND(G) 163L, Health and Safety Executive

HSE (1995) *Approved code of Practice for Safety Data Sheets*, L62, Health and Safety Executive

HSE (1997) *The Costs of Accidents at Work*, Health and Safety Executive

Hester, R and Harrison, R (1997) *Contaminated Land and its Reclamation RSC, Issues in Environmental Science and Technology No: 7*

Hill, T (1999) 'It's a dirty business', *Estates Gazette*, 9901, Jan 9 1999, pp108–109
Hovey, J (1997) 'Expanding Environmental Market has Huge Unrealised Potential' *Rough Notes*, March 1997, pp26–28

ICA (1998) *Financial Reporting of Risk*, The Institute of Chartered Accountants 1998.

ICA (1999) *Internal Control: Guidance for Directors on the Combined Code* The Institute of Chartered Accounts in England and Wales

IEM (1998) 'Credit Rating for the Environment', *Industrial Environmental Management*, March 1998, pp28–29

Lawrence, C M, Fortune, M and Butler, A C (1995) 'A Fuzzy Decision Model' for Environmental Risk Assessment in Studies in Managerial and Financial Account, M J Epstein Vol. 3 *Applications of Fuzzy Sets and the Theory of Evidence to Accounting,* Jai Press Inc, pp213–236

Lindgaard-Jorgensen, P and Bender, K (1992) *Review of Environmental Accidents and Incidents,* Community Documentation Centre on Industrial Risk EUR14002

Lockett, N (1996) *Environmental Liability Insurance*, Cameron May

Loose, HJE (1998) 'Environmental Insurance: The New Panacea or Merely a Placebo?' in *Business Law Review*, August/September 1998 pp194–199

Marshall, G (1997) Towards holistic risk management: following Cadbury, who will lead and how? *AIRMIC Newsletter*, May 1997, pp8–11

Merkl, A and Robinson, H (1997) 'Environmental Risk Management: Take it Back from the Lawyers and Engineers.' *The McKinsey Quarterly*, 3, pp151–163

Metelski, J J and Gulledge, W P (1989) 'Providing EIL Coverage' in Calabrese, E (ed) *Petroleum Contaminated Soils* Vol 2 Lewis

Nathanail, P (1995) *Environmental Threats to Human Health, Eco-systems and Built Structures*, Association for Geographic Information, AGI Conference Proceedings

Ouellette, R (1990) 'Check, please' in *Site Assessment 1990 Management Sourcebook,* Environmental Publications Inc, Lutherville.

Parry, J (1998) Spains Bitter Spill, *International Risk Management*, September 1998, pp30–31

Pearson, C R (1988) 'Assessment of risk of significant environmental contamination from storage and distribution sites' in *IChemE Symposium series,* 110, pp105–130.

Petts, J (1994) 'Contaminated Sites: Blight, Public Concerns and Communication' *Land Contamination and Reclamation*, 2, 4, pp171–181

POST (1996) *Safety in Numbers: Risk Assessment in Environmental Protection*, June

Power M and McCarty, L (1998) 'A Comparative Analysis of Environmental Risk Assessment/Risk Management Frameworks' *Environmental Science and Technology*, May, pp224A–231A

Pritchard, P (1998) 'Environmental Living' *Corporate Insurance and Risk*, July 1998,p18

Priznar, F J (1991) 'Environmental audit types and their appropriate use', in *Environmental Auditing in Cleaner Production Strategies International Expert Seminar,* Trolleholm Castle, Sweden April 18–19, Dept. Industrial Environmental Economics, Lund University.

Ramsey, C (1998) 'Constructing an integrated approach to managing safety, environment and quality' *Journal of the Institution of Occupational Safety and Health* 2, 2, pp25–43

Richards, T (1997) *Is it worth the risk? The Impact of Environmental Risk on Property Investment Valuation*, The College of Estate Management, Reading

RICS (1997) *Contamination and its Implications for Chartered Surveyors:* A Guidance Note, Royal Institution of Chartered Surveyors

Roche, D (1996) 'Landfill failure survey: A technical note' in Bentley, SP (ed) *Engineering Geology of Waste Disposal.* Geological Society Engineering Geology Special Publication No 11, pp379–380

Rodricks, J (1994) *Calculated Risk,* Cambridge University Press

The Royal Society (1992) *Risk: Analysis Perception and Management* (1992) Study Group Report

Schroeder, P (1990) 'Risk assessment issues associated with managing existing hazardous waste sites' in *Integrating Risk Management for Hazardous Wastes* M M Kunreuther and R Gowda, Kluwer, pp105–130.

Smets, H (1988) 'The cost of accidental pollution' in *UNEP Industry and Environment,* October/December, pp28–32.

Steward, M (1996) 'Environmental Risk Management and Finance' *Journal of Property Development* 1, 2, pp56–59

Swartjes, F A (1997) *Assessment of Soil and Groundwater Quality in the Netherlands: Criteria and Remediation Priority*, WasteTECH Symposium, 17-19 March 1997, Melbourne, Australia

Syms, P (1996) 'Property professionals and their perceptions' *Estates Gazette*, October 26 1996, pp146–148

Syms, P (1997) *Contaminated Land: The Practice and Economics of Redevelopment*, Blackwell Science

Taylor, J (1995) *Testimony of Jerry Taylor*, Congressional Testimony, June 22 1995

Thompson, K-M (1992) 'Monte Carlo techniques for Quantitative Uncertainty Analysis in Public Health Risk Assessments' in *Risk Analysis*, 12, 1, pp53–63

Turner, NJK, Gronow, SA and Pritchard PW (1994) 'Assessing the Environmental Risks of Property Investment Portfolios' *Journal of Property Finance*, 5, 4, pp68–83

Tusa, W (1994) 'Towards a Definition of Environmental Risk Management' *Risk Management*, December, pp 60–61

Upton, P S (1998) *The Use of Risk Assessment and Geostatistics in the Management of Contaminated Land*, Proceedings of MANENPRO, 3 March 98. The Society of Chemical Industry

USEPA (1997) *Guidelines for Monte Carlo Analysis* (EPA/630/R-97/001) US Environmental Protection Agency

Waring, A (1998) 'Risk Profiling for the 21st century', *Risk Management Bulletin* July/August 1998, Vol 3, I, pp4–9

Waring, A (1998b) 'Complacent about COMAH: then wake up to FRR' *The Chemical Engineer*, 24 September 1998, p 21

Waring, A and Glendon, A (1998) *Managing Risk*, International Thomson Business Press

Winter, M and Steger, U (1998) *Managing Outside Pressure: Strategies for Preventing Corporate Disasters,* John Wiley

Woellner, R (1998) 'ISO 14001 as a tool for implementing environmental risk management programmes' *Strategies and Solutions 98*, London, 8 October 1998

Wright, A (1998) 'Banks: Managing Credit Risks and Minimising Environmental Liability' *Strategies and Solutions 98* Conference - 8 October, London

Writer, N (1998) 'The 1995 Environment Act: A threat to property owners and their insurers' *Journal of the Society of Fellows* (CII) Vol 12, Pt 2, pp3–20, 1998

Yorkshire and Humberside Regional Technology Network (1994) *Contaminated Land Report,* June 1994

APPENDIX A – ENVIRONMENTAL ISSUES IN CORPORATE TRANSACTIONS

A Survey on Issues and Outcomes

In the UK many organisations first experienced environmental liability issues when involved in corporate transactions with US counterparties. The US experience with increasing regulatory requirements through the 1970s and 1980s, in particular Superfund, raised concern about the inheritance of other people's liabilities. This led to the employment of specialists to undertake environmental work as part of the due diligence process. A survey of leading UK practitioners conducted by Royal & SunAlliance in 1998 is described here to give an indication of what might be expected when commissioning an environmental survey as part of the due diligence process.

Survey Methodology

A questionnaire was distributed to a group of the most prominent UK providers of environmental consultancy services. This approach was intended to permit openness, in that corporate purchasers or vendors would not need to disclose information that related directly to their own activities. In addition, a large sample size could be accessed relatively easily. The returns related to a sample of just over 1,900 recent UK audits.

A common feature is the phased approach including Phase 1 type studies involving a desk review of available records supplemented by a site visit from an environmental specialist, on to Phase 2 intrusive studies involving the removal of samples for laboratory analysis. Time and cost were the principal drivers behind this phased approach. Detailed investigations of ground conditions can be both expensive and time consuming, usually requiring several weeks to complete. In the context of fast moving corporate transactions this length of delay can be detrimental. Similarly, a large proportion of sites can often be excluded from detailed investigation by simple screening to demonstrate there are unlikely to be significant liabilities at the properties.

Several respondents noted that consultants are often commissioned when the deal is at an advanced stage, resulting in significant time pressures. Clarification of terms of reference as soon as is practicable appears to be the message coming back to the procurers of such services.

Potential misunderstandings arising out of the scope of the work and other related issues are likely to mean a receptive audience for an international standard. The International Standards Organisation is currently considering ISO 14015 – Guidelines for the Assessment of Sites and Entities. In terms of differences between studies in the UK and overseas, the general difficulty in obtaining historical data (eg on site operations) outside the UK was highlighted.

A number of factors were said to have raised the prominence of environmental issues in property transfer in the UK:

- *Contaminated land:* the current revision of the regulatory regime, including questions of assigning liability as part of the Environment Act 1995. It was felt (in early 1998) that the

Environment Act was already exerting a general influence in raising awareness of potential liabilities associated with contaminated land.

- *Legislation such as the Integrated Pollution Control regime (and the forthcoming IPPC Directive):* requiring organisations to consider compliance costs associated with improvement programmes.

- *Better understanding of historical liabilities through implementation of formal environmental management systems such as ISO 14001.* Although ISO 14001 registration can be maintained by two companies in the same sectors with very different liability profiles, it usually ensures that much better information is available to facilitate the due diligence process.

Who Commissions the Work?

It is unsurprising that the work was most commonly commissioned by prospective purchasers, accounting for almost half of the total (49 per cent), as compared to 5 per cent coming from the vendor. In some cases the vendor requests may arise as a response to the findings from the purchaser's advisers. The potential for disagreement (and delay inducing argument) can lead in some cases to a joint commission from both parties. The 'caveat emptor' principle would also seem to be pertinent to the desire to avoid inheritance of liabilities by purchasers.

The incorporation of environmental risk management criteria into banking operations would seem to be reflected by the fact they were responsible for commissioning over a quarter (26 per cent) of the total. By way of comparison a US study found investors and lenders constituted 12 per cent of the total. Advisers such as lawyers were found to account for 16 per cent of the demand in the US.

What were the Outcomes?

The most likely outcome following the audits was that the deal proceeded, either with no significant problems identified (27 per cent) or with some account taken of environmental issues, eg an amended price or warranty (33 per cent). In almost one-quarter of the cases the identified potential for significant liabilities was such that a Phase 2 study was initiated. In a small, but nonetheless, significant number of cases (4 per cent) the results were considered to have been a significant factor in the prospective purchaser backing away from the deal. It is interesting to note that 75 per cent of the audits in the US study revealed no apparent problem, a significantly higher figure than the UK. This might reflect the more routine nature of the undertaking in the US, whereas the studies have historically only been triggered in the UK where potential problems are already anticipated. It was also stated that large companies usually have a good general understanding of environmental risk management, which coupled with the size of potential liabilities (particularly outside the US), means that deals are rarely jeopardised. Problematic properties or operations may be ring fenced and excluded from the deal, or warranties could be offered by the selling parties.

Phase 2 intrusive studies aim to confirm and quantify, as appropriate, the extent of potential liabilities highlighted by the Phase 1 work. Of the total Phase 2 project outcomes 3 per cent led to the breakdown of the transaction, while 30 per cent indicated that the potential issues identified earlier were not significant. The most frequent result was that the deal progressed with some account taken of the Phase 2 findings – this is unsurprising given that the financial implications of liabilities should have become much clearer. In 19 per cent of cases detailed Phase 3 investigations were instigated. Phase 3 work was often said to be undertaken as part of site redevelopment, the planning process providing the main avenue for regulatory involvement in contaminated land.

Outcomes of environmental audits

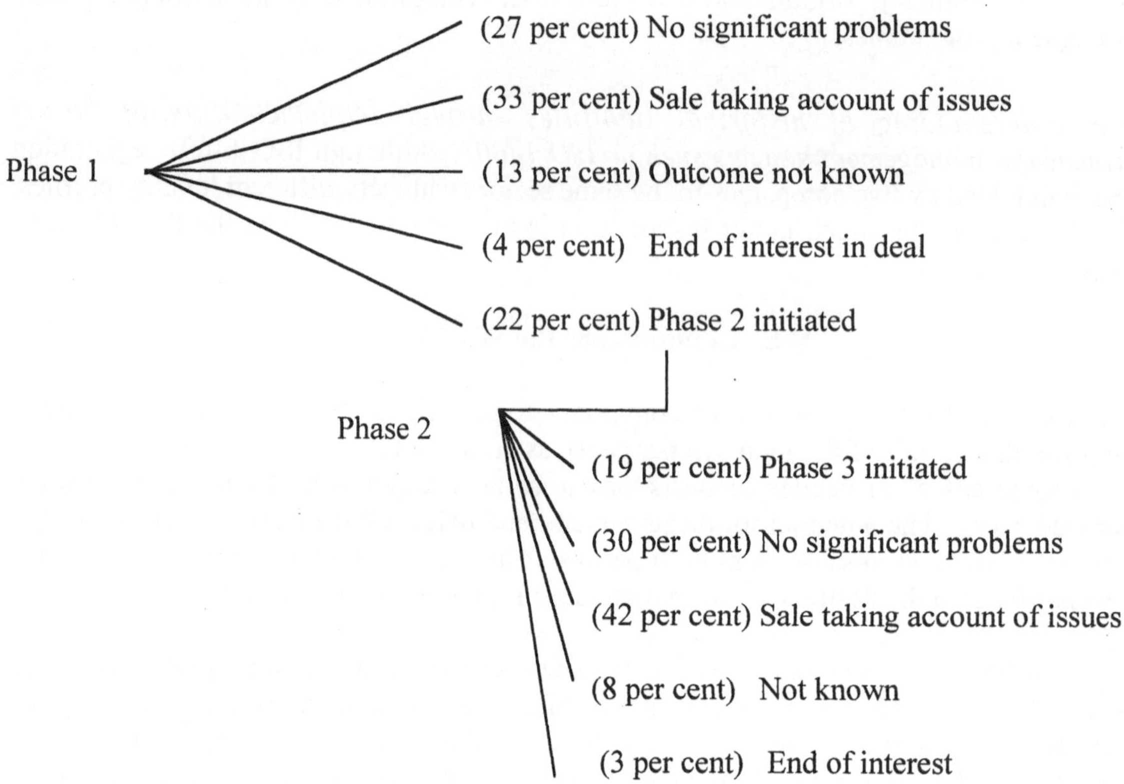

What are the Main Issues?

Contaminated land remains the principal concern, however, respondents also said upgrade costs to ensure ongoing legal compliance of plant and process equipment had become more important. Other potentially significant concerns included:

- the presence of asbestos on site;
- noise problems, which can prove difficult and expensive to abate;
- the presence of toxic polychlorinated biphenyls (PCBs), most notably in electrical transformer oil;
- location in an area which has elevated levels of naturally occurring radon; and
- underground storage tanks – where leaks of oil, fuel etc. may go unnoticed for many years.